The
Mathematical
Universe

An Alphabetical Journey Through the Great Proofs, Problems, and Personalities

William Dunham

John Wiley & Sons, Inc.
New York ■ Chichester ■ Weinheim ■ Brisbane
Singapore ■ Toronto

This book is dedicated to Brendan and Shannon

This text is printed on acid-free paper.

Copyright © 1994 by John Wiley & Sons, Inc.
All rights reserved. Published simultaneously in Canada.
First published in 1994 by John Wiley & Sons, Inc.

The poem "Misunderstanding," which appears on page 164,
is © 1993 by JoAnne Growney.

Library of Congress Cataloging-in-Publication Data:

Dunham, William
    The mathematical universe : an alphabetical journey through the great
proofs, problems, and personalities / William Dunham.
      p. cm.
    Includes index.
    ISBN 0-471-53656-3 (cloth)
    ISBN 0-471-17661-3 (paper)
    1. Mathematics—History. 2. Mathematicians—History.   I. Title.
QA21.D785   1994
510—dc20   93-48720

Printed in the United States of America

10 9 8 7 6 5 4 3 2 1

Contents

PREFACE

Many children begin reading with an alphabet book. Comfortably seated on a warm lap, youngsters listen as the alphabet unwinds from "A is for alligator" to "Z is for zebra." Such books may not be great literature, but they provide an effective introduction to letters, words, and language.

Echoing those alphabet readers of childhood, this volume surveys the discipline of mathematics in a series of essays running from A to Z. The content is considerably more sophisticated—D now stands for differential calculus rather than doggie—and a warm lap is no longer mandatory. But the basic idea of an alphabetical journey remains.

Such a format imposes severe restrictions upon a book meant to be read from cover to cover. Mathematical topics, after all, do not align themselves in a logical progression so as to mirror the Latin alphabet. Consequently, the transition between chapters is sometimes abrupt. Moreover, although certain letters of the alphabet are rich in possible subjects, others are quite barren, a situation reminiscent of children's alphabet books in which "C is for cat" but "X is for xenurus." As the reader will readily note, some of the following topics have been shoehorned into position, much as a size 16 foot must be squeezed into a size 8 boot. Devising a topical flow to coincide with the alphabetical flow has presented a true logistical challenge.

The book begins with the (apparently) simple subject of arithmetic. Later chapters have recurring and often interlaced themes. Sometimes consecutive chapters fit together, as with Chapters G, H, and I on geometry or the back-to-back Chapters K and L on seventeenth-century rivals Isaac Newton and Gottfried Wilhelm Leibniz. Some chapters focus on single mathematicians: We meet Euler in Chapter E, Fermat in Chapter F, and Bertrand Russell in Chapter R. Some describe specific results, such as the isoperimetric problem or Archimedes' determination of the surface area of a sphere. Some address such broader issues as the mathematical personality or the presence of women in the discipline. And whatever the subject, each chapter provides a strong dose of history.

Along the way, the major branches of mathematics—from algebra, to geometry, to probability, to calculus—make at least a brief appearance. Those sections designed to explain key mathematical ideas have the flavor of an informal textbook, and actual proofs (or at least "prooflets") appear here and there. Chapters D and L, for instance, introduce differential and integral calculus, and thus carry a bit more mathematical baggage.

In most chapters, however, there has been a deliberate attempt to avoid an overtly technical development. Virtually all of the mathematics is elementary—that is, pitched to those with some high school algebra and geometry under their belts. Professional mathematicians will find few surprises on these pages. The book is aimed at those whose interest in mathematics is at least as extensive as their training.

A few themes return again and again: that mathematics is an ancient yet vital subject, that it treats matters of everyday importance as well as matters of no utility whatever, and that it is a discipline whose remarkable breadth is matched only by its equally remarkable depth. To convey something of this in a sequence of alphabetically ordered chapters is the book's objective.

I would be remiss not to mention John Allen Paulos's book *Beyond Numeracy* (Knopf, New York, 1991), which he described as "in part a dictionary, in part a collection of short mathematical essays, and in part the ruminations of a numbers man." Paulos's lively volume charted a mathematical course from A to Z—in his case, from algebra to Zeno. By allowing multiple entries for some letters, he achieved greater breadth of coverage; by writing fewer but longer essays, I opted for greater depth. It is my hope that our two books can coexist peacefully as variations of the same alphabetical format.

Of course, there is no way an author can address every key point, introduce every important figure, or consider every pressing mathematical issue. Choices must be made at each turn, choices constrained by the demands of internal consistency, by the complexity of the subject, by the author's interest and expertise, and by the utterly artificial arrangement of the alphabet. A project of this type will omit a thousand times more than it can include, and scores of potential topics are bound to end up on the cutting room word processor.

In the end, this book is the response of a single individual to the immense mathematical universe. It represents one of countless journeys that could have been undertaken by countless authors, and I make no claim to having followed the comprehensive or definitive route from A to Z.

Such qualifications aside, I hope these chapters provide at least a glimpse into a subject of endless fascination. As the nineteenth-century mathematician Sofia Kovalevskaia observed, "Many who have never had the occasion to discover more about mathematics confuse it with arithmetic and consider it a dry and arid science. In reality, however, it is a science which demands the greatest imagination."[1] And perhaps the book can serve to underscore Proclus' lofty sentiment from fifth-century Greece: "mathematics alone can revive and awaken the soul . . . to the vision of being, can turn her from images to realities and from darkness to the light of the intellect."[2]

ACKNOWLEDGMENTS

In the course of writing this book, I have received the support of friends and relatives, of colleagues and editors. Among these it is important to single out a few.

Special thanks are due to Daryl Karns, who first suggested the idea of an alphabet book about mathematics. Daryl is one of the great biology professors, a liberal artist extraordinaire, and, I am pleased to say, a dear friend.

As a new faculty member at Muhlenberg College, I have greatly appreciated the welcome extended by President Arthur Taylor and by my colleagues in the Department of Mathematical Sciences: John Meyer, Bob Stump, Roland Dedekind, Bob Wagner, George Benjamin, and Dave Nelson. The same appreciation goes out to the staff of Muhlenberg's Trexler Library for their cheerful assistance during the preparation of this manuscript.

Beyond Muhlenberg, I thank colleagues Don Bailey, Victor Katz, Alayne Parson, and Buck Wales for assistance at various points in the preparation of the manuscript. And I gratefully acknowledge my editors at John Wiley & Sons: Steve Ross, who served as midwife in bringing this book into existence, and Emily Loose and Scott Renschler, who carried it through adolescence to maturity.

An especially warm thank you goes to my mother, to Ruth and Bob Evans, and to Carol Dunham for their constant love and encouragement.

Most of all, I thank my wife and colleague, Penny Dunham. She was instrumental in determining the book's contents and in outlining some of its chapters. As maestro of the Macintosh, she created the figures and diagrams contained within. And her editing of the manuscript improved the final product immeasurably. Without question, Penny's influence is everywhere evident on the pages that follow.

W. Dunham
Allentown, Pennsylvania, 1994

rithmetic

For each of us, mathematics begins with arithmetic, and so does this book. As we all know, arithmetic deals with the most basic of quantitative concepts, the whole numbers 1, 2, 3 . . . . If any mathematical idea is universal, it is that of distinguishing degrees of multiplicity, which is to say "counting."

The intrinsic inevitability of the whole numbers, their undeniable naturalness, lies at the heart of Leopold Kronecker's well-known observation, "God made the integers; all else is the work of man."[1] If we imagine mathematics as a grand orchestra, the system of whole numbers could be likened to a bass drum: simple, direct, repetitive, providing the underlying rhythm for all the other instruments. There surely are more sophisticated concepts—the oboes and French horns and cellos of mathematics—and we examine some of these in later chapters. But whole numbers are always at the foundation.

Mathematicians call the infinite collection, 1, 2, 3, . . . , the positive integers, or perhaps more descriptively, the natural numbers. Once we have introduced and named them, our attention turns to combining them in certain important ways. Most fundamental is addition. This operation is not only fundamental, but it is also *natural* in the sense that the numbers are created additively. That is, $2 = 1 + 1$, $3 = 2 + 1$, $4 = 3 + 1$, and so on. Even as a powerful thoroughbred is "born to run," so, too, are the natural numbers "born to add."

During elementary-school days, we add numbers *ad infinitum* (almost) and then take up the opposite, or inverse, operation of subtraction. Next, it is on to multiplication and division, accompanied by seemingly interminable periods of drill. After many years of this instruction, youngsters master the operations of arithmetic with varying degrees of imperfection, undaunted by the fact that a $7.95 calculator can do

1

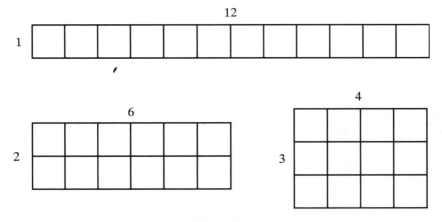

Figure A.1

it all flawlessly in a fraction of the time. It is unfortunate that, to so many young people, arithmetic becomes synonymous with drill and drudgery.

In the not-too-distant past, however, the term *arithmetic* embraced not only the basic operations of addition, subtraction, multiplication, and division but also the deeper properties of the whole numbers. Europeans, for instance, spoke of the "higher arithmetic," by which was meant, in all honesty, the "harder arithmetic." The preferred term nowadays is *number theory*.

This subject, though ranging far and wide, is more or less anchored on the notion of prime numbers. A whole number greater than 1 is **prime** if it cannot be written as the product of two smaller whole numbers. Thus the first ten primes are 2, 3, 5, 7, 11, 13, 17, 19, 23, and 29. None of these has a positive integer divisor other than 1 and itself.

An argumentative reader might contend that 17 *can* be written as a product—for instance, $17 = 2 \times 8.5$ or $17 = 5 \times 3.4$. But in these cases the factors are not themselves integers. It must be remembered that in number theory the starring roles are played by whole numbers; their more sophisticated and far-ranging cousins—the fractions, the irrationals, the imaginaries—must remain tantalizingly offstage.

If a whole number greater than 1 is not prime—which is to say, if a number possesses an integer factor other than 1 and itself—we say it is **composite**. Numbers such as $24 = 4 \times 6$ or $51 = 3 \times 17$ are examples. For reasons that will become apparent shortly, the whole number 1 is excluded from the ranks of either primes or composites. Consequently the smallest prime is 2.

A simple and frequently cited method of visualizing these concepts is to imagine square pieces of floor tile that must be assembled into a rectangular shape. If we have 12 such pieces, we enjoy different options as to the rectangle that can be formed, as shown in Figure A.1. This, of course, is because $12 = 1 \times 12$, or $12 = 2 \times 6$, or $12 = 3 \times 4$ (we do not distinguish between 3×4 and 4×3 because the shape of the resulting floor is the same in both cases, although one is rotated). Like-

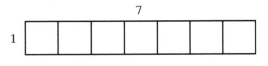

Figure A.2

wise, 48 pieces of tile yield five different floor plans, corresponding to the decompositions $48 = 1 \times 48 = 2 \times 24 = 3 \times 16 = 4 \times 12 = 6 \times 8$.

On the other hand, there is but one floor plan we can make out of 7: the obvious and not very interesting 1×7 shown in Figure A.2. If someone must tile a rectangular room with exactly 7 pieces of flooring, he or she had better have a very long, narrow room in mind. From this perspective, a number p is prime if it has but one plan: the trivial $p = 1 \times p$. A number is composite if its floor plan has options.

Prime numbers may lie at the heart of the higher arithmetic, but they also are responsible for its greatest mathematical snarls. The reason is simple: Whereas the whole numbers are literally created by the operation of addition, questions about primes and composites thrust *multiplication* into the picture. Number theory is so difficult, albeit so fascinating, because mathematicians try to examine additive creations under a multiplicative light.

In this sense, the natural numbers resemble fish out of water. Spawned by the process of addition, they find themselves in an unfamiliar, multiplicative environment. Of course, before we dismiss the whole enterprise as hopeless, we should recall that 350 million years ago fish really *did* come out of water; took a few inefficient breaths in a world for which they were imperfectly designed; and went on to evolve into amphibians, reptiles, birds, mammals, and mathematicians. Sometimes a new, hostile environment can make all the difference.

Primes certainly would occupy a less central position in number theory were it not for a result known as the fundamental theorem of arithmetic (note the use of the term *arithmetic* in its broader sense). As the name suggests, this is one of the most basic, most important propositions in all of mathematics. It is stated simply:

FUNDAMENTAL THEOREM OF ARITHMETIC: Any positive integer (other than 1) can be written as the product of prime numbers in one and only one way. ■

Here we have a two-edged sword asserting, first, that any whole number can be expressed as the product of primes in *some* fashion and, second, that there is but *one way* to do this. We are led inevitably to the conclusion that primes are the multiplicative building blocks from which all whole numbers are assembled, and herein lies their importance. Primes play a role analogous to that of the chemical elements, for just as any natural compound can be broken into a combination of the 92 natural elements on the periodic chart (or the 100+ elements including those created in the laboratory), so, too, can any whole number be decomposed into its prime factors. A molecule of the compound we call water, H_2O, can be separated into two atoms of the element hydrogen and one atom of the element oxygen. Similarly, the compound (i.e., composite) number 45 can be broken into a product of two factors of the prime 3 and one of the prime 5. Mimicking water's chemical notation, we could write $45 = 3_25$, although mathematicians prefer the exponential form $45 = 3^2 \times 5$.

But arithmetic's fundamental theorem provides more than just a decomposition into primes. Equally critical is its guarantee of the uniqueness of such decompositions. If someone determines the prime factorization of 92,365 to be $5 \times 7 \times 7 \times 13 \times 29$, then a colleague—working across the room or across the country, working today or a thousand centuries from now—*must* come up with precisely the same prime decomposition.

This is very comforting to mathematicians. Of course, it is equally comforting that, when a chemist decomposes a water molecule into one oxygen and two hydrogen atoms, a different chemist is not going to decompose that molecule into one atom of lead and two of molybdenum. Primes, like elements, are not only building blocks but also unique building blocks.

It is worth noting that the desire for unique factorization requires us to exclude 1 from among the primes. For if 1 were classified as a prime, then the number 14, for instance, would have prime decomposition $14 = 2 \times 7$ as well as the *different* prime decompositions $14 = 1 \times 2 \times 7$ and $14 = 1 \times 1 \times 2 \times 7$. The uniqueness of prime factorization would vanish. It is far better, say mathematicians, to give 1 a special role. Neither prime nor composite, it is called a ***unit***.

Confronted with a positive integer, a mathematician may wish to determine whether it is prime or composite and, in the latter case, to find its prime factors. Sometimes the matter is simple. Any even whole number (other than 2) is obviously not prime because it has a factor of 2, and any whole number ending in the digits 5 or 0 (other than 5) is likewise composite. In other cases, the question of primality presents far greater difficulties. Who, for instance, would care to determine which of 4,294,967,297 and 4,827,507,229 is prime and which is not?*

The nineteenth-century mathematician Carl Friedrich Gauss (1777–1855), perhaps the greatest number theorist of all, put it simply in his 1801 masterpiece *Disquisitiones Arithmeticae*:

> The problem of distinguishing prime numbers from composite numbers and of resolving the latter into their prime factors is known to be one of the most important and useful in arithmetic.... The dignity of the science itself seems to require that every possible means be explored for the solution of a problem so elegant and so celebrated.[2]

For more than two dozen centuries, from ancient Greeks to modern number theorists, mathematicians have been irresistibly drawn to such problems, even as moths are drawn to a flame or spaghetti sauce to a white shirt. Along the way scholars have made a multitude of conjectures about the primes. Some have been proved, others disproved, and a surprising number remain unresolved.

---

* Believe it or not, 641 divides evenly into 4,294,967,297; the other number is prime. See David Wells, *The Penguin Dictionary of Curious and Interesting Numbers*, Penguin, New York, p. 192.

For instance, an intriguing question was raised by the French priest Marin Mersenne (1588–1648) in 1644. Mersenne played an important role in seventeenth-century science, not only for his contributions to the theory of numbers but also for his valuable service as a clearinghouse of information among mathematicians. Scholars curious about the state of mathematics or perplexed by a difficult problem would write Mersenne, who either knew the answer or could direct them to a likely authority. The value of such a channel of information—in the days before scientific meetings, professional journals, or e-mail—can hardly be overestimated.

Mersenne was intrigued by numbers of the form $2^n - 1$; that is, numbers that are one less than a power of 2. Today these are called **Mersenne numbers** in his honor. Obviously, all such numbers are odd. More important, some of them are prime.

Immediately Mersenne recognized that if n is composite, then $2^n - 1$ must be composite, too. For instance, if $n = 12$, the Mersenne number $2^{12} - 1 = 4095 = 3 \times 3 \times 5 \times 7 \times 13$ is composite (because 12 is); for the composite $n = 33$, $2^{33} - 1 = 8,589,934,591 = 7 \times 1,227,133,513$ is likewise not a prime.

When the exponent is prime, however, the situation is less clear. Letting $p = 2, 3, 5$, and 7 yields the "Mersenne primes" $2^2 - 1 = 3$, $2^3 - 1 = 7$, $2^5 - 1 = 31$, and $2^7 - 1 = 127$. But if we use the prime $p = 11$ as exponent, we get $2^{11} - 1 = 2047$; alas, this number is the product of 23 and 89 and is therefore composite. Mersenne was fully aware that p being prime is no guarantee that $2^p - 1$ is prime as well. In fact, he asserted that the only primes between 2 and 257 for which $2^p - 1$ is prime are $p = 2, 3, 5, 7, 13, 17, 19, 31, 67, 127$, and 257.[3]

Unfortunately, Father Mersenne's assertion contained sins of both commission and omission. For instance, he missed the fact that the number $2^{61} - 1$ is a prime. On the other hand, $2^{67} - 1$ turned out not to be prime at all. This latter fact was established in 1876 by Edouard Lucas (1842–1891), who demonstrated that the number was composite using an argument so indirect that it did not explicitly exhibit any of the factors. In a sense, then, the story of $2^{67} - 1$ remained less than complete, and its final chapter is worth a brief digression.

The year was 1903 and the setting was a meeting of the American Mathematical Society. Among the speakers on the program was Frank Nelson Cole of Columbia University. When his turn came, Cole walked to the front of the room, silently multiplied 2 by itself 67 times, subtracted 1, and arrived at the monumental result 147,573,952,588,676,412,927. Having witnessed this wordless computation, the perplexed spectators next watched as Cole wrote on the board

$$193,707,721 \times 761,838,257,287$$

which he also calculated in silence. The product was none other than

$$147,573,952,588,676,412,927$$

Cole sat down. His performance would have been perfectly appropriate at a convention of mimes.

Those in the audience—who had just witnessed the explicit factorization of Mersenne's number $2^{67} - 1$ into two gigantic factors—were momentarily as speechless as Cole. Then they burst into unrestrained applause and gave him a standing ovation! This recognition, it is hoped, warmed Cole's heart, for he later admitted that he had been working on his calculation for the previous two *decades*.[4]

Cole's factorization notwithstanding, Mersenne numbers remain a fruitful source of primes. Almost certainly, when a newspaper proclaims that a new "largest" prime has been found, it turns out to be of the form $2^p - 1$. As of 1992, the largest known prime was $2^{756839} - 1$, a behemoth of 227,832 digits.[5] But the general question of determining which Mersenne numbers are prime and which are not remains an unsolved problem of number theory.

The Mersenne number $2^7 - 1 = 127$ figures in another tale about primes. In the mid-nineteenth century, the French mathematician A. de Polignac asserted:

Every odd number can be expressed as the sum of a power of 2 and a prime.[6]

For instance, 15 can be written as $8 + 7 = 2^3 + 7$, whereas $53 = 16 + 37 = 2^4 + 37$ and $4107 = 4096 + 11 = 2^{12} + 11$. Although de Polignac did not claim to have proved his fascinating conjecture, he implied he had checked it for all odd numbers up to three million.

Because any power of 2 has no odd number in its prime decomposition, such powers are somehow the most purely even of all numbers. De Polignac's statement suggested that any odd number could be built out of a prime, that most fundamental of building blocks, and a purely even power of 2. It was a bold statement.

It was also dead wrong. If de Polignac really spent the time necessary to check his assertion into the millions, we can only pity him, for the relatively tiny Mersenne prime 127 refutes his claim; there is no way to write 127 as a power of 2 plus a prime. This is evident if we simply decompose 127 in all possible ways into a power of 2 and a remainder and observe that the latter is never prime:

$$127 = 2 + 125 = 2 + (5 \times 25)$$
$$127 = 4 + 123 = 2^2 + (3 \times 41)$$
$$127 = 8 + 119 = 2^3 + (7 \times 17)$$
$$127 = 16 + 111 = 2^4 + (3 \times 37)$$
$$127 = 32 + 95 = 2^5 + (5 \times 19)$$
$$127 = 64 + 63 = 2^6 + (3 \times 21)$$

(Because $2^7 = 128$ is larger than 127, we need go no further.) De Polignac's conjecture is now relegated to the number theoretic scrap heap, for he overlooked a counterexample that was, in mathematical terms, right under his nose. Like the ornithopters of the nineteenth century, his ambitious assertion never got off the ground.

We have drawn parallels between the unique decomposition of chemicals into elements and the unique factorization of integers into primes. This chemical analogy, although helpful, breaks down in one significant way: All the laboratory work of all the chemists in history has expanded the supply of elements to just over 100, but the collection of primes is infinite. Although the periodic chart of chemical elements can fit on a medium-size wall, a similar chart of mathematical primes would require a wall of endless extent.

The earliest proof of the infinitude of primes is due to the Greek mathematician Euclid (ca. 300 B.C.) and appeared in his classic work, the *Elements*.[7] We present a slightly modified version of his argument, but one that retains the power and beauty of the original.

To follow the reasoning requires two preliminary results from number theory, neither of them very deep. The first is that for any whole number n, the difference between two multiples of n is itself a multiple of n. Symbolically, if a and b are multiples of n, so is $a - b$. For example, both 70 and 21 are multiples of 7, and so is their difference, $70 - 21 = 49$; likewise, both 216 and 72 are multiples of 9, as is $216 - 72 = 144$. The general proof of this fact is not included here, but its verification is as simple as it is believable.

The other prerequisite is equally elementary. It says that any composite number has at least one prime factor. Again we illustrate by example. The composite 39 has the prime factor 3, the composite 323 has the prime factor 17, and the composite 25 has the prime factor 5. Euclid gave a clever proof of this theorem as the 31st proposition of the seventh book of the *Elements*.

Beyond that, all that is necessary to establish the infinitude of primes is an understanding of proof by contradiction. This in turn requires that we recognize the most basic dichotomy of logic: A statement is either true or false.

One way to demonstrate the truth of a statement is to prove it directly. This is obvious (and banal). A different but equally valid approach—the so-called ***proof by contradiction***—is to assume the statement is *false* and from this assumption, using the rules of logic, derive an impossible consequence. The appearance of such a consequence indicates that something is wrong somewhere along the chain of reasoning, and if our steps were valid the only possible culprit is the original assumption of the statement's falsehood. We must therefore reject this falsehood, and the dichotomy mentioned above leaves us with a single possibility: The statement must in fact be true. Admittedly, this strategy may seem strangely indirect and unnecessarily devious. To emphasize the indirectness, we consider the following example before returning to the infinitude of primes.

Suppose we are investigating numbers that are both perfect squares and perfect cubes, as 64 is both 8^2 and 4^3 or 729 is both 27^2 and 9^3. Such a number will be called a "sqube." Our goal is to prove:

THEOREM: There are infinitely many squbes.

PROOF: This is easy and perfectly direct. We merely observe that if n is any whole number, then $n^6 = n^3 \times n^3 = (n^3)^2$ is a perfect square, and $n^6 = n^2 \times n^2 \times n^2 = (n^2)^3$ is simultaneously a perfect cube. So we get an infinitude of squbes by looking at

$$1^6 = 1^2 = 1^3$$
$$2^6 = 64 = 8^2 = 4^3$$
$$3^6 = 729 = 27^2 = 9^3$$
$$4^6 = 4096 = 64^2 = 16^3$$
$$5^6 = 15625 = 125^2 = 25^3$$
$$6^6 = 46656 = 216^2 = 36^3$$
$$7^6 = 117649 = 343^2 = 49^3$$
$$8^6 = 262144 = 512^2 = 64^3$$ and so on

Obviously this process continues without end, for each different choice of n generates a new and different n^6. The infinitude of squbes has been demonstrated *directly*.
■

Unfortunately, to prove the infinitude of primes we have no such direct option. Neither Euclid nor anyone since has found a simple formula that churns out primes as our formula n^6 churned out squbes. Rather than make a frontal assault, we must resort to the indirect attack of proof by contradiction—more sophisticated, more subtle, but in the end far more beautiful. In fact, this proof often serves as a kind of litmus test for mathematical sensibility: Those with a genuine mathematical yen find it moves them to tears; those without such a yen find it bores them to tears. We let the reader be the judge.

THEOREM: There are infinitely many primes.

PROOF: (by contradiction) Assume instead that there are only finitely many primes, and suppose they are denoted by $a, b, c, \ldots, d$. This collection may contain 400 primes or 400,000, but we assume it contains them *all*. We now start marching toward a contradiction.

Multiply these primes together and add 1 to create the new number:

$$N = (a \times b \times c \times \ldots \times d) + 1$$

Note that because we have only finitely many numbers, we can indeed multiply them in this fashion; an infinitude of primes could not have been so multiplied.

Obviously N is larger than any of the individual primes $a, b, c, \ldots,$ or d, so N is different from all of them. Since these were the only primes in town, we conclude that N is not one of the primes.

This means N must be composite, and by the second of our preliminary observations above, N has a prime divisor. Because we assumed that $a, b, c, \ldots, d$ constitute the world's entire supply of primes, this prime divisor of N must be somewhere among them.

Put another way, N is a multiple of one of the primes $a, b, c, \ldots,$ or d. It really doesn't matter which prime it is, but for the sake of concreteness, assume N is a multiple of c. Clearly the product $a \times b \times c \times \ldots \times d$ is also a multiple of c because c appears as one of the factors. By the first preliminary cited above, the difference between N and $a \times b \times c \times \ldots \times d$ will also be a multiple of c. But we defined N to be just 1 more than this product, so the difference is 1.

We thus have been led to conclude that 1 is a multiple of c (or of any other prime that is a factor of N). This is clearly impossible because the smallest prime is 2 and thus 1 is not a multiple of *any* prime. Something has gone haywire.

As we scan back through the argument, we see the only source of trouble is our original assumption that there are only finitely many primes. We therefore reject this and conclude, by contradiction, that the primes must be infinitely abundant. The proof is complete. ■

This wonderful piece of reasoning is elementary yet profound. It guarantees that primes are endless and inexhaustible. After the most powerful computer confirms that $2^{756,839} - 1$ is prime, we can smugly assert that even bigger primes—indeed, infinitely many bigger primes—still lay undiscovered. Even if we are unable to specify any of these larger primes, let no one think we are being evasive. Thanks to the subtleties of logic and proof by contradiction, we know they are out there.

Because it boasts results of such simple beauty, number theory has served as the point of entry into higher mathematics for many young scholars. Among these was the U.S. mathematician Julia Robinson (1919–1985). In 1970 Robinson was among a trio of scholars who solved what is known as Hilbert's Tenth Problem, a very deep question from number theory that had remained unanswered since it was posed by David Hilbert (1862–1943) seven decades earlier. Even as a youngster, Robinson had been enthralled by the wonderful properties of the whole numbers. "I was especially excited by some of the theorems of number theory," she wrote, "and I used to recount these to Constance [her sister] at night after we went to bed. She soon found that if she wasn't ready to go to sleep she could keep me awake by asking questions about mathematics."[8]

Or there is the Hungarian mathematician Paul Erdös (pronounced "air-dish"). Reminiscing on his long career, Erdös (1913–) recalled, "When I was ten, my father told me about Euclid's proof [of the infinitude of primes], and I was hooked."[9]

Erdös's youth was as intellectually productive as it was socially sheltered. At the age of 17, when most college freshmen are simply hoping to survive adolescence, Erdös achieved mathematical fame by devising an elementary proof that between any whole numbers n and $2n$ there must always lie at least one prime. For instance, there must be a prime between 8 and 16, or between 8 billion and 16 billion.

This may seem like a fairly unspectacular theorem. Indeed, it had been proved almost a century earlier by the Russian mathematician with the mellifluous name of Pafnuty Lvovich Chebyshev (whose appearance in the mathematical literature as Chebychev, Tchebysheff, Cebysev, and Tshebychev should be attributed to errors of transliteration rather than a fondness for aliases). But Chebyshev's proof was very complicated. The amazing thing about Erdös's argument was that it was so much simpler and came from one so young.

We note in passing that his theorem provides an alternate proof of the infinitude of primes, for it guarantees that there is a prime between 2 and 4, a prime between 4 and 8, another between 8 and 16, and so on. Even as we can keep doubling numbers forever, so, too, must the primes be limitless.

This was the first of a long line of theorems from Paul Erdös, the twentieth century's most prolific, and perhaps most eccentric, mathematician. Even in a profession in which unusual behavior is accepted as something of the norm, Erdös is legendary. For instance, so sheltered was this young scholar that only at the age of 21—four years *after* proving the theorem on primes mentioned above—did he first butter his own bread. He later reminisced: "I had just gone to England to study. It was tea time, and bread was served. I was too embarrassed to admit that I had never buttered it. I tried. It wasn't so hard."[10]

Equally unusual is that Erdös has no permanent residence. Instead, he travels around the globe from one mathematical research center to another, living out of a suitcase and trusting that at each stop someone will put him up for the night. As a result of his incessant wanderings, this vagabond mathematician has collaborated with more colleagues, and published more joint papers, than anyone in history. He is proof of the scriptural adage that a human being does not live by (buttering) bread alone.

The mathematical community, in return, has devised a whimsical recognition of his impact: the so-called Erdös number.[11] Erdös himself has Erdös number 0; any mathematician who has jointly published a paper with Erdös has Erdös number 1; a mathematician who, though not collaborating directly with Erdös, has jointly published a paper with someone who published a paper with Erdös receives Erdös number 2; someone who published a paper with someone who published a paper with someone who published a paper with Erdös has Erdös number 3; and so on. Like a gigantic oak, the Erdös tree has branched across the mathematical world.

So, with prime numbers, composite numbers, Mersenne numbers, and even Erdös numbers, it is clear that the passion for number theory is in no danger of subsiding. For mathematicians from Gauss to Robinson, from Euclid to Erdös, no part of mathematics has proved more beautiful, elegant, or endlessly fascinating than the higher arithmetic.

ernoulli Trials

At the outset, we stress that a Bernoulli trial is not a Florentine legal proceeding. Rather, it is a cornerstone of the elementary theory of probability and thus plays an important role in understanding our uncertain world.

A **Bernoulli trial** is simply an experiment with a dichotomous outcome. It results in either success or failure, black or white, on or off. There is no middle ground, no room for compromise, no comfort for the wishy-washy.

Examples abound. We deal a card from a deck and classify it as either black or red. We give birth to a child and get either a girl or a boy. We go through a 24-hour day and either do or do not get hit by a meteor. In each case it is convenient to designate one outcome as the "success" and the other as the "failure." For instance, choosing a black card, having a daughter, and not getting struck by a meteor could be labeled successes. From a probabilistic perspective, however, it makes no difference whether the red card, the son, or the meteor are regarded as successes. In this arena, the word *success* carries no value judgment.

A single Bernoulli trial is rarely of much interest. The plot thickens, however, when Bernoulli trials are conducted repeatedly, and we observe how many of these trials yield successes and how many yield failures. This accumulated record can shed some very useful information on the underlying processes.

One stipulation is essential as we conduct our experiments: The repeated trials must be **independent**. This term has a technical definition but also carries an informal meaning adequate for our purposes: Namely, two events are independent if the outcome of one has absolutely no impact on the outcome of the other. Thus, for instance, the birth of a son to the Smiths and that of a daughter to the Johnsons are independent events. So, too, are the outcomes (heads or tails) of tossing a dime and a penny; the outcome of one coin has no bearing on that of the second.

But if we deal two cards from a deck, one at a time, and regard a black card as a success, independence is lost in going from the first card to the second. For if the first is the ace of clubs (a success), this will influence the outcome on the next draw, making it somewhat less likely to be black. It also is somewhat less likely to be an ace and absolutely certain not to be another ace of clubs.

Fortunately this lack of independence can be remedied by a simple expedient. After drawing the first card, we put it back into the deck, shuffle well, and draw again. Because our first card has been replaced within a thoroughly mixed deck, its identity can have no impact upon our next draw. In this sense, independent events require a clean slate for each experiment so that the probability of a success remains the same from trial to trial.

The clearest examples of Bernoulli trials arise in games of chance, such as the tossing of coins or the rolling of dice. For coins there is a clear independence from flip to flip so that the probability of a success—such as getting a head—is the same on each toss. By saying a coin is "balanced," we mean that this probability is exactly 1/2. For a fair die, if we designate a success as rolling a 3, our probability of success is always 1/6.

But what happens if we flip a coin five times? What is the probability that we get a mix of three heads and two tails among the five flips? For that matter, if we flip the coin 500 times, what is the chance of getting 247 heads and 253 tails? This may seem like a nightmarish problem, but its solution appears in one of the early master-pieces of probability theory, the *Ars Conjectandi* by Jakob Bernoulli (1654–1705).

Bernoulli was a native of Switzerland whose grandfather, father, and father-in-law were all prosperous pharmacists. Forsaking the mortar and pestle, Jakob studied theology in college, obtaining a degree at the age of 22. But though his family was in serums and his training in sermons, Jakob Bernoulli's real interest lay in mathematics.

From the late 1670s until the end of his life, Bernoulli was one of the world's leading mathematicians. His remarkable talent was coupled with a prickly personality, a massive ego, and a tendency to belittle the efforts of those less gifted. For instance, after studying what are now called (in his honor) the "Bernoulli numbers," Jakob found an ingenious shortcut for summing the powers of positive integers. He observed that "it took me less than half of a quarter of an hour" to determine the sum of the tenth powers of the first 1,000 positive integers. That is, in less than ten minutes he determined

$$1^{10} + 2^{10} + 3^{10} + 4^{10} + \ldots + 1,000^{10}$$
$$= 91,409,924,241,424,243,424,241,924,242,500$$

a truly stupendous sum. But in a self-serving editorial comment, Jakob pointed out that his shortcut made it "clear how useless was the work of Ismael Bullialdus . . . in which he did nothing more than compute with immense labor the sums of the first

Jakob Bernoulli
(Reprinted by permission of Birkhäuser Verlag AG, Basel, Switzerland, from
Jakob Bernoulli, Collected Works, Vol. 1: Astronomia, Philosophia naturalis;
edited by Joachim O. Fleckenstein, 1969)

six powers, which is only a part of what we have accomplished in the space of a single page."[1] One's heart goes out to poor Ismael, who had as critic a mathematician not only of enormous insight but also of enormous ego.

Jakob Bernoulli's productive years coincided with Gottfried Wilhelm Leibniz's discovery of calculus, and Jakob was one of the chief popularizers of this immensely fruitful subject. As with any developing theory, calculus benefited from those who followed in its creator's footsteps, scholars whose brilliance may have fallen short of Leibniz's but whose contributions toward tidying up the subject were indispensable. Jakob Bernoulli was one such contributor.

In this undertaking, he had an uneasy ally in Johann (1667–1748), the younger of what we might call, with unashamed alliteration, the brilliant but bickering Bernoulli

Figure B.1

brothers. Jakob, in fact, had played a role in teaching mathematics to his younger sibling. In later years he probably regretted that he taught Johann so well, for the youngster turned out to be a mathematician to rival, if not surpass, his teacher. There arose between the brothers a fierce competition for mathematical supremacy. Johann expressed unconcealed glee when he solved a problem that had stumped his brother, whereas Jakob demeaningly called Johann his "pupil," implying that Johann was only parroting his mentor's brilliance. Neither Bernoulli was what one would call gracious.

A famous clash came with the problem of the catenary. A catenary is the curve assumed by a hanging chain affixed to a wall at two points (see Figure B.1). Those familiar with high school algebra may guess that the chain hangs in a parabolic arc, a perfectly reasonable conjecture advanced by no less a figure than Galileo earlier in the seventeenth century. But the hanging chain is *not* parabolic, and by 1690 Jakob Bernoulli was working hard to determine the curve's true identity—that is to say, its equation.

As it turned out, Jakob was not up to the task. It is easy to imagine his consternation when Johann came forth with the answer. Johann later gloated over his triumph, asserting that the solution "cost me study that robbed me of rest for an entire night."[2] As tactless as he was brilliant, Johann rushed to Jakob and presented the correct solution to his still perplexed brother. Jakob was crestfallen.

But Jakob would have his revenge. This time the battleground was the so-called isoperimetric problem. The issue at hand was to identify, from among all curves with the same perimeter, that which encloses the *maximum* area. We examine this more closely in Chapter I, but for now we observe that Jakob Bernoulli applied calculus to a version of this problem in 1697. His work required him to wrestle with a complicated mathematical entity called a third-order differential equation and pointed the way to a new, important, and far-reaching branch of mathematics now known as the calculus of variations.

Brother Johann disagreed and claimed to have resolved the isoperimetric challenge with a considerably simpler *second-order* differential equation. As was so often the case in the house of Bernoulli, their disagreement decayed into antagonism and stopped just short of musket fire.

Johann Bernoulli
(Courtesy of Carnegie Mellon University Library)

This time, however, it was Jakob who earned the last laugh, for his brother's second-order differential equation was incorrect. Unfortunately, Jakob never actually
got a chance to laugh, or even grin, for he died in 1705 while Johann's incorrect
solution to the problem remained inexplicably sealed in the offices of the Paris
Academy. There is speculation that Johann had realized his error and arranged to

conceal the mistake so as not to endure a public humiliation while his brother was around to enjoy it.[3]

These episodes give a sense of the friction that characterized this brotherly relationship. It is little wonder that, when Johann was suggested as the obvious person to edit the papers of his recently deceased brother, Jakob's widow forbade this, fearing that the vindictive Johann would sabotage Jakob's mathematical legacy.[4] Perhaps J. E. Hofmann put it best in his characterization of Jakob in the *Dictionary of Scientific Biography:* "He was self-willed, obstinate, aggressive, vindictive, beset by feelings of inferiority, and yet firmly convinced of his own abilities. With these characteristics, he necessarily had to collide with his similarly disposed brother."[5] Indeed, Jakob and Johann Bernoulli are the kind of people who give arrogance a bad name.

Sibling rivalry aside, we return to the probabilistic question posed earlier: If a balanced coin is tossed five times, what is the probability it yields three heads and two tails? In the *Ars Conjectandi* Jakob Bernoulli gave a general rule: If we conduct $n + m$ repeated, independent trials of an experiment (that is, $n + m$ Bernoulli trials), where the probability of a success on any trial is p and the probability of a failure is $1 - p$, then the chance of getting exactly n successes and m failures is given by the formula

$$\frac{(n + m) \times (n + m - 1) \times \ldots \times 3 \times 2 \times 1}{[n \times (n - 1) \times \ldots \times 3 \times 2 \times 1] \times [m \times (m - 1) \times \ldots \times 3 \times 2 \times 1]} \, p^n (1 - p)^m$$

To simplify this expression, mathematicians introduce the *factorial* notation:

$$n! = n \times (n - 1) \times \ldots \times 3 \times 2 \times 1$$

For instance, $3! = 3 \times 2 \times 1 = 6$ and $5! = 5 \times 4 \times 3 \times 2 \times 1 = 120$. (We stress that the exclamation point in the factorial notation does *not* require us to speak in a loud voice.) With this notational convention, Bernoulli's result simplifies to

$$\text{Prob } (n \text{ successes and } m \text{ failures}) = \frac{(n + m)!}{n! \times m!} p^n (1 - p)^m$$

Thus the probability of getting three heads on five flips of a balanced coin is found by letting $n = 3$, $m = 2$, and $p = \text{Prob (flipping a head)} = 1/2$. That is,

$$\text{Prob (3 heads, 2 tails)} = \frac{(3 + 2)!}{3! \times 2!} \left(\frac{1}{2}\right)^3 \left(1 - \frac{1}{2}\right)^2 = \frac{5!}{3! \times 2!} \left(\frac{1}{8}\right)\left(\frac{1}{4}\right)$$

$$= \frac{120}{6 \times 2} \left(\frac{1}{32}\right) = 0.3125, \text{ or just over } 31\%$$

Likewise, to find the probability of getting exactly five 4s on 15 independent rolls of a die, we declare getting a 4 to be a "success" and assign the values:

$$n = 5 \text{ (number of successes)}$$
$$m = 15 - 5 = 10 \text{ (number of failures)}$$
$$p = 1/6 \text{ (probability of success)}$$

so that the chance of getting five 4s on 15 rolls is

$$\frac{(5+10)!}{5! \times 10!}\left(\frac{1}{6}\right)^5\left(1-\frac{1}{6}\right)^{10} = \frac{15!}{5! \times 10!}\left(\frac{1}{6}\right)^5\left(\frac{5}{6}\right)^{10} = 0.0624$$

a fairly unlikely happening.

And, to return to an earlier question, the chance of getting 247 heads and 253 tails on 500 flips of a coin is

$$\frac{(247+253)!}{247! \times 253!}\left(\frac{1}{2}\right)^{247}\left(1-\frac{1}{2}\right)^{253} = \frac{500!}{247! \times 253!}\left(\frac{1}{2}\right)^{247}\left(\frac{1}{2}\right)^{253}$$

Although correct, this probability is much too complicated to determine by hand, and even an expensive pocket calculator will balk at an attempt to compute something like 500! (skeptics are invited to try it). We shall see a technique of approximating such probabilities in Chapter N. But even if the direct computation is forbidding, the formula is theoretically flawless. It is the key to finding probabilities for any string of independent Bernoulli trials.

Unfortunately, most events from everyday life turn out to be more complicated than the tossing of coins, a probabilistic situation that is almost too pure. Far less accessible is the problem of determining the probability that a 25-year-old man will live to age 70 or beyond, or the probability that it will rain more than an inch next Tuesday, or the chance that a car approaching an intersection will turn right. These events are contaminated by the immense complexity of the real world, as Jakob recognized when he wrote:

> what mortal, I ask, could ascertain the number of diseases, counting all possible cases, that afflict the human body in every one of its parts and at every age, and say how much more likely one disease is to be fatal than another—plague than dropsy, for instance, or dropsy than fever—and on that basis make a prediction about the relationship between life and death in future generations?[6]

Are such probabilities beyond the realm of mathematics? Is probability theory to be relegated merely to artificial games of chance?

Bernoulli provided a powerful answer to this question in a result from *Ars Conjectandi* that may be his greatest legacy. In fact, he called it his "golden theorem" and wrote, "Both its novelty and its great utility, coupled with its just as great difficulty, exceed in weight and value all the other chapters of this doctrine."[7] Known today as Bernoulli's theorem or, more commonly, as the law of large numbers, it stands as one of the central pillars of the theory of probability.

To understand something of its nature, again suppose we are conducting independent Bernoulli trials, where the probability of a success is p on each trial. We keep track of the total number of trials performed—call this N—as well as the number of these trials that result in a success—call this x. Then the fraction x/N is simply the *proportion* of times we observe a success.

For instance, if 100 flips of a balanced coin produce 47 heads, the observed proportion of heads is $47/100 = 0.47$. If the coin is flipped another 100 times and yields another 55 heads, the combined proportion of successes is

$$\frac{47 + 55}{100 + 100} = \frac{102}{200} = 0.510$$

And there is nothing (short of boredom) to stop someone from tossing the coin another hundred times, or another million times for that matter. The key question is, what happens to x/N, the proportion of successes, over the long run?

No one should be too surprised to find this proportion approaching ever closer to 0.5 as the number of repetitions of the experiment increases. In general, as N grows larger, we observe the changing value of x/N tending toward the fixed number p, the *true* probability of a success on any individual trial. So—and here lies the theorem's power—when the probability of a success p is *unknown*, the proportion of successes among a large number of trials (x/N) should provide a good estimate of p. In symbols, we write

$$\frac{x}{N} \approx p \text{ if } N \text{ is large}$$

(where $\approx$ means "is approximately equal to").

This, with a few important qualifications, is the law of large numbers. What is notable about Jakob Bernoulli's theorem is not that it is true, but that it was so difficult to prove with a rigorous argument. Jakob himself acknowledged, in his typically uncharitable manner, that "even the stupidest man knows [the law of large numbers] by some instinct of nature."[8] Yet a valid proof of this law cost him 20 years of effort and fills pages of the *Ars Conjectandi*.[9] His comment that "the scientific proof of this principle is not at all simple" proved to be a decided understatement.

We should say a word about the aforementioned "important qualifications" regarding Bernoulli's theorem. Because it is essentially a probabilistic statement, it is subject to the uncertainties that come with any chance occurrence. We cannot be absolutely positive that, upon tossing a coin a thousand times, the proportion of heads will be nearer to 0.5 than the proportion resulting from a mere 100 flips. It is conceivable that 51 of the 100 flips result in heads but that only 486 of the 1000 flips land heads up. Thus, the "small sample" estimate $x/N = 51/100 = 0.51$ actually would be closer to the true probability of tossing a head than is the "large sample" estimate $x/N = 486/1000 = 0.486$. Chance can do such things.

For that matter, it is not absolutely impossible that, if we tossed the coin an additional 1000 times, each of these flips would come up heads. This would result in an astounding 1486 heads on 2000 flips, and an estimated probability of $1486/2000 = 0.743$. In such a case, the law of large numbers seems to have been repealed.

But not really. For what Jakob Bernoulli proved was that, given any small tolerance—say 0.000001—the *chance* that the estimated probability x/N and the true probability p differ by this tolerance or less can be made as close to 1 as we like merely by increasing the number of trials. By taking enough trials, we can be almost certain—Bernoulli used the term *morally certain*—that our estimate x/N is within 0.000001 of the true probability p.[10] Admittedly, we are not 100 percent confident that p and x/N agree within 0.000001, but the large number of trials makes us certain enough of this agreement so as not to lose any sleep over the matter.

The situation above in which the probability of flipping a head was estimated as 0.743 after 2000 tosses of a balanced coin is less likely than the chance of a person's getting hit with a meteor while reading this chapter. Moreover, even if such an improbable estimate arose, Bernoulli could smugly assert that by conducting more trials—2000 more, or 2000 million more—the ratio of x/N would, with "moral certainty," return to the vicinity of 0.5.

It is important to stress that, even with these fairly minor qualifications, the law of large numbers is *provable*. This distinguishes it from other famous laws encountered in life, from Murphy's law to the law of gravity. These are either generally accepted platitudes (as with the former) or highly regarded physical models (as with the latter) that are always subject to modification should new evidence arise. But the law of large numbers is a mathematical *theorem*, proved within the unforgiving constraints of logic to stand forever.

And it certainly has its uses. Survival probabilities governing the actuarial tables of insurance companies are based upon the results of a large number of similar trials (i.e., of people's lives and deaths). So, too, are the weather forecaster's chances of rain.

Or consider the example, dating back to the eighteenth century, of finding the probability that a woman gives birth to a boy rather than a girl. How could anyone calculate this probability in an a priori fashion? Complex factors of genetics hopelessly confuse the situation of determining beforehand in some purely theoretical way the chance of having a boy. Instead, we are forced to launch an "after-the-fact," or a posteriori, attack with our weapon being Bernoulli's law.

This very issue was on the mind of John Arbuthnot of England in the early years of the 1700s. Like others before him, he noted from census records that slightly more boys than girls were born each year and asserted that this imbalance has been present "for ages of ages, and not only at London, but all over the world."[11] Arbuthnot tried to account for this phenomenon by invoking a divine intervention in human affairs. Some years later, Nicholas Bernoulli, nephew of Jakob and Johann in this talented mathematical family, applied the law of large numbers to conclude that the probability of bearing a boy was $18/35$. In other words, the large number of recorded births showed a marked and stable tendency toward 18 males for every 17 females. Bernoulli's theorem had been put to work "not only at London but all over the world."

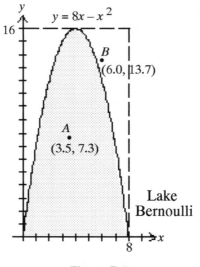

Figure B.2

It is still at work today. A technique known as the Monte Carlo method, which combines Bernoulli's theorem with the power of the computer, has become quite important in allowing scientists to simulate a wide range of random phenomena in a probabilistic fashion. The following is a simplistic yet illustrative example of the Monte Carlo method. Suppose we wish to find the surface area of an irregularly shaped lake. We could walk its borders or take an aerial photograph, but the lake's curving and seemingly random boundaries tend to defy any simple mathematical formula for the determination of its area.

We suppose our lake is mapped in Figure B.2, where we have superimposed the standard x and y axes onto a diagram. Because we plan to return to this example in Chapter L, we have chosen a lake that is quite tame—one bounded by the x-axis and the parabolic segment with equation $y = 8x - x^2$.

We shall estimate its area probabilistically. First, enclose the region within the 8×16 rectangle shown in the diagram. Next, unleash the computer to select at random hundreds of (x, y) points within this rectangle. For instance, the computer might churn out points $A = (3.5, 7.3)$ or $B = (6.0, 13.7)$ as shown.

We now ask the computer whether these random points fall inside the lake or not. In our example, this question is easily resolved. To check point A, we let $x = 3.5$ in the parabola's equation and find the associated value of $y = 8(3.5) - (3.5)^2 = 15.75$. This means that the point $(3.5, 15.75)$ lies *on* the parabola. Consequently A, with the same first coordinate but with second coordinate of only 7.3, falls below the parabola and lands in the water.

Similarly, when considering point B, we substitute its first coordinate into the equation and get a value of $y = 8(6) - 6^2 = 12$. Hence $(6, 12)$ is on the parabola, so $B = (6, 13.7)$ falls a bit above the curve and thus hits dry land. With a few milliseconds of computer time, we can choose scores of random points and determine whether or not they fall into the lake.

But now comes the critical observation from the Monte Carlo viewpoint: The *exact* probability a randomly chosen point falls within the lake—denoted by p—is precisely the proportion of the 8×16 rectangular area that the lake occupies. That is,

$$p = \text{Prob (a random point lands in the lake)} = \frac{\text{area of lake}}{\text{area of the enclosing rectangle}}$$

$$= \frac{\text{area of lake}}{8 \times 16} = \frac{\text{area of lake}}{128}$$

Of course, we cannot calculate this probability unless we know the lake's area, the very unknown we seek. But we can *estimate* the probability p of hitting the lake by x/N, the proportion of hits in the shaded region. This use of the long run proportion of successes to approximate the true probability of success is a direct application of the law of large numbers.

For this example, our computer selected 500 points in the rectangle and found that 342 of them hit the lake. Thus, we estimated

$$\frac{342}{500} \approx p = \frac{\text{area of lake}}{128}$$

which after cross-multiplication becomes

$$\text{area of lake} \approx 128 \times \frac{342}{500} = 87.552 \text{ square units}$$

We thereby have a crude approximation of the lake's size without recourse to anything but Bernoulli's theorem and a computer.

How could we get a sharper estimate? We simply let the computer select not 500 but 5000 points at random within the rectangle. In this case, it found 3293 of these landing in the water, yielding

$$\frac{3293}{5000} \approx p = \frac{\text{area of lake}}{128}$$

and so

$$\text{area of lake} \approx 128 \times \frac{3293}{5000} = 84.301 \text{ square units}$$

Of course, we could now ask the computer for 50,000 random points, or 500,000, or as many as our electric bill permits. With ever greater confidence, we would obtain estimates of the area of the parabolic lake.

This is an elementary and artificial example, and real-world phenomena of much more sophistication can be explored via Monte Carlo methods. Moreover, as we shall see later, the parabolic area in question can be determined *exactly* by means of integral calculus. Still, this example gives a sense of the power of probability.

Three centuries have passed since Jakob Bernoulli proved his great theorem. As is not uncommon in mathematics, his original argument since has been replaced by a streamlined version that cuts to the heart of the matter much more efficiently. Today's standard proof depends upon a result due to the Russian mathematician Pafnuty Chebyshev, whom we met in Chapter A. This approach, involving such concepts as the expected value and standard deviation of a random variable, allows us to condense the proof of the law of large numbers into just a paragraph and makes Bernoulli's multipage argument appear cumbersome indeed. However, in a spirit of generosity unknown to Jakob we shall resist the temptation to label his work "useless" because he required a chapter to do "what we have accomplished in the space of a single page."

Such is the nature of progress. But, as in all human endeavors, it is best to remember the pioneers. Just as today's laser disk technology provides recorded music infinitely superior to the scratchy sound of a nineteenth-century phonograph, so, too, has modern probability theory shortened and simplified Bernoulli's proof of the law of large numbers. Yet we still revere Thomas Edison in spite of subsequent advances that have rendered his original creation obsolete. It is only fitting that we accord the same respect to Jakob Bernoulli for the golden theorem of which he was so justly proud.

Circle

$C/D = \pi$

The first two chapters introduced the domains of number theory and probability. Before proceeding further, we should consider a topic from geometry, one of the major branches of mathematics. Geometry was the primary focus of the Greek mathematicians, as we shall see in Chapter G, and has a rich and distinguished history. So prominent was this subject in the classical world that the words *mathematician* and *geometer* were synonymous. To a great extent, geometry was what mathematicians did.

One can of course introduce geometry from many different angles. The present chapter examines the circle, surely one of the most important of geometric concepts. Circles are simple, elegant, and beautiful, exhibiting a genuine two-dimensional perfection. In the hands of the Greeks they were of interest not only in themselves but also as primary tools in the exploration of other geometric ideas.

The terminology of circles has become part of our common vocabulary. By definition a *circle* is a plane figure all of whose points lie the same distance from a fixed point. The fixed point is called the *center,* and the common distance of all points from the center is the *radius.* The distance across the circle through the center is the circle's *diameter,* and the length of the circular curve itself—the distance one would travel in making a complete circuit—is the *circumference.*

A novice, introduced to circles for the first time, would rather quickly recognize an essential fact: All circles have the same shape. Some may be large and some small, but their "circleness," their perfect roundness, is immediately evident. Mathematicians say that all circles are *similar*. Before dismissing this as an utterly trivial observation, we note by way of contrast that not all triangles have the same shape,

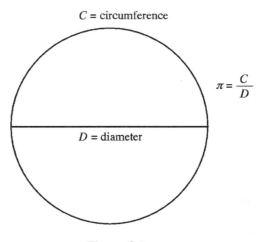

Figure C.1

nor all rectangles, nor all people. We can easily imagine tall narrow rectangles or tall narrow people. But a tall narrow circle is not a circle at all.

So, circles are all the same shape. Behind this unexciting observation lies one of the profound theorems of mathematics: that the *ratio* of circumference to diameter is the same for one circle as for any other. Whether the circle is gigantic, with large circumference and large diameter, or minute, with tiny circumference and tiny diameter, the *relative* size of circumference to diameter will be the exactly the same. Letting C stand for circumference and D for diameter, mathematicians say the ratio C/D is constant from one circle to another.

What do we call this constant? Never missing an opportunity to introduce a new symbol, mathematicians chose the sixteenth letter of the Greek alphabet, π, instantly elevating it to a kind of mathematical immortality. The choice is fitting because the Greeks were the first to subject the circle to mathematical scrutiny, but the Greeks themselves did not use π in this sense.

To formalize the concept, we consider Figure C.1 and introduce:

DEFINITION: If C is a circle's circumference and D is its diameter, then $C/D = \pi$.

When cross-multiplied, this definition yields the famous formula $C = \pi D$. Alternately, because the diameter is twice the radius, we restate this to get the equally well-known $C = 2\pi r$.

Thus π provides the critical connection between circumference (a length) and radius (another length). It is immensely important, although not intuitively clear, that this same constant does an equally good job connecting the circle's *area* and radius. It is worth a moment's discussion to see why this is so.

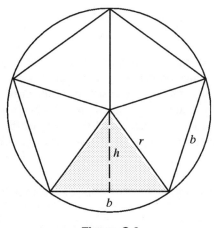

Figure C.2

The critical idea is to approximate a circle with an inscribed **_regular polygon,_** that is, a polygon with sides all the same length and angles all the same measure. Polygons are more accessible figures than circles, yet our knowledge of polygons can lead us to an understanding of the circles within which they lie.

In Figure C.2 we see a regular pentagon inscribed in a circle of radius r. To determine the pentagon's area, we draw radii from the circle's center to the five vertices on the circle, thereby breaking the pentagon into five triangular pieces. Each triangle has base of length b, the side of the pentagon, and height h, the dotted line drawn perpendicularly from the center of the circle to the side of the polygon and called the **_apothem._** By the well-known formula for triangular area, we see that

$$\text{area(each triangle)} = \frac{1}{2}\,(\text{base})\,(\text{height}) = \frac{1}{2}\,bh$$

and so

$$\text{area(inscribed pentagon)} = 5 \times \text{area (each triangle)} = 5 \times \left(\frac{1}{2}\,bh\right) = \left(\frac{1}{2}\,h\right) \times 5b$$

But $5b$ is just five times the length of the pentagon's side and consequently is the pentagon's perimeter. In short, we have established that

$$\text{area(regular pentagon)} = \left(\frac{1}{2}\,h\right) \times \text{perimeter}$$

A moment's thought reveals that the same formula holds whether we inscribe within the circle a regular 5-gon or 20-gon or 1000-gon. For the general case, in which is inscribed a regular n-gon, the polygon will be divided into n little triangles, each with apothem h (the distance from the center of the circle perpendicular to the side) and base b (the length of a side of the n-gon). Hence

$$\text{area(polygon)} = n \times \text{area (triangle)} = n \times \left(\frac{1}{2} bh\right) = \left(\frac{1}{2} h\right) \times nb = \left(\frac{1}{2} h\right) \times \text{perimeter}$$

because the perimeter is n times the length b of each side.

Now imagine inscribing consecutively a regular 10-gon, a regular 10,000-gon, a regular 10,000,000-gon, and so on, increasing the number of sides without end. It is clear, at least intuitively, that in this fashion the polygons will gradually "fill up"—the Greeks said "exhaust"—the circle, and the areas of the inscribed figures will approach the circular area as an upper limit. Using the notation *lim* to stand for *limit,* we see that

$$\text{area (circle)} = \lim\,[\text{area (regular inscribed polygons)}] = \lim\left[\left(\frac{1}{2} h\right) \times \text{perimeter}\right]$$

Never will an inscribed polygon have *exactly* the same area as the circle, for straight sides, no matter how tiny, will never coincide precisely with a circular arc. Nonetheless, the polygonal areas will approach the limiting area of the circle as closely as one wishes.

Two questions remain: What happens to the apothem and to the perimeter as the number of polygonal sides increases indefinitely? Clearly h will have as its limit the radius of the circle. Likewise the limiting value of the perimeters of the inscribed regular n-gons will be the circle's circumference. These facts can be stated symbolically as

$$\lim h = r \quad \text{and} \quad \lim\,(\text{perimeter}) = C$$

Hence

$$\text{area(circle)} = \lim\left[\left(\frac{1}{2} h\right) \times \text{perimeter}\right] = \left(\frac{1}{2} r\right) \times C = \frac{rC}{2}$$

At last π makes its appearance, for we noted above that $C = \pi D = 2\pi r$. The preceding formula thus becomes:

$$\text{area(circle)} = \frac{rC}{2} = \frac{r(2\pi r)}{2} = \frac{2\pi r^2}{2} = \pi r^2$$

This is without doubt one of the key formulas in all of mathematics, one that resonates not only with mathematicians but even with newspaper cartoonists (see example, opposite).

So, if we want to find either the circumference or the area of a given circle, we are sure to encounter π. But this raises a practical problem, namely, to determine the value of this key ratio. After all, π is only a symbol for a real, honest-to-goodness number, and anyone having to do a calculation with circles needs to know that number (at least approximately). One can no more find a circle's numerical area using only the *symbol* π than one can bake a cake using only the word *egg*.

(FRANK & ERNEST, reprinted by permission of NEA, Inc.)

The simplest way to approximate the ratio C/D is to measure the circumference and diameter of a particular circle and divide the former by the latter. For instance, a piece of string wrapped around the outside of a bicycle tire measured 82 inches, whereas another piece stretched diametrically across the tire came out at 26 inches. Our real-world experiment thus yielded the estimate: $\pi = C/D \approx 82/26 = 3.15\ldots$, where $\approx$ stands for "is approximately equal to," as it did in the previous chapter. Unfortunately, when a similar procedure was applied to the circumference and diameter of the circular top of a coffee can, we got $\pi = C/D \approx 18/6 = 3.00$, which is not terribly close to the first estimate. Physical measurements such as these obviously introduce inaccuracies, and in any case tangible coffee cans or bicycle wheels are not perfect, mathematical circles.

For a purely mathematical estimate of the ratio of circumference to diameter, we turn to Archimedes of Syracuse (287–212 B.C.), a revered figure from the history of mathematics. Absentminded, self-absorbed, and somewhat eccentric, Archimedes was recognized as a scientific genius even in his own time. For better or worse, he is probably best remembered today for the incident with the crown of King Hieron.

According to legend, the king of Syracuse had given a craftsman a certain amount of gold to be made into an exquisite crown. When the project was completed, a rumor surfaced that the craftsman had substituted a quantity of silver for an equivalent amount of gold, thereby devaluing the crown and defrauding the king. Was the rumor correct? The task of uncovering the truth was assigned to Archimedes. We pick up the story with the words of the Roman architect Vitruvius:

> While Archimedes was considering the matter, he happened to go to the baths. When he went down into the bathing pool he observed that the amount of water which flowed outside the pool was equal to the amount of his body that was immersed. Since this fact indicated the method of explaining the case, he did not linger, but moved with delight, he leapt out of the pool, and going home naked, cried aloud that he had found exactly what he was seeking. For as he ran he shouted in Greek: *eureka, eureka*![1]

Although of questionable authenticity, this is certainly a famous story. Probably no other tale in all of science combines the elements of brilliance and bareness quite so effectively.

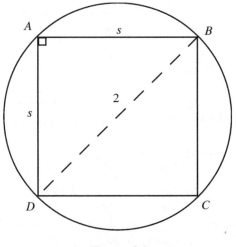

Figure C.3

Historians report that Archimedes often did his mathematics by drawing figures in the sand. It is even said that he carried a portable sand tray, the laptop computer of the time. When inspiration struck, he would place the tray upon the ground, smooth out the sand, and begin to draw his geometric diagrams. From today's perspective, such a medium offers clear disadvantages: A strong burst of wind could blow away a brilliant proof; a bully could kick a theorem into your face; and should a cat wander into the tray, the outcome could be too disgusting to contemplate.

Yet Archimedes prevailed, creating a body of mathematics that left not only his contemporaries but also generation after generation of scholars in awe. We return to him in Chapter S, where we examine in some detail one of his greatest triumphs, the determination of the surface area of a sphere. But here the focus is on his splendid estimate of the ratio of circumference to diameter—in other words, his approximation of π.

Archimedes' strategy, like that above, approached the circle via regular polygons. The following development, although employing modern notation and a slightly different starting point, is consistent with that strategy. It requires only a bit of algebra and the Pythagorean theorem. The latter, of course, states that in a right triangle the square on the hypotenuse equals the sum of the squares on the other two sides. (A discussion of the theorem of Pythagoras is available to anyone who wishes to turn to Chapter H.)

We refer to Figure C.3, where we have inscribed square $ABCD$ in a circle. Because the ratio of circumference to diameter is the same for all circles, we might as well make our job easier by choosing the radius of this circle to be $r = 1$. Thus the square's diagonal, the dotted line in the figure, is the circle's diameter, $2r = 2$.

We denote by s the length of each side of the square, so that right triangle ABD has two sides of length s and hypotenuse of length 2. From the Pythagorean theorem it follows that $s^2 + s^2 = 2^2$, so that $2s^2 = 4$ and $s = \sqrt{2}$. Consequently the perimeter of the square is $P = 4s = 4\sqrt{2}$.

The square's perimeter provides a first, albeit rough, estimate of the circle's circumference. Replacing circumference by perimeter, we get

$$\pi = \frac{\text{circumference}}{\text{diameter}} \approx \frac{\text{perimeter}}{\text{diameter}} = \frac{4\sqrt{2}}{2} = 2\sqrt{2} = 2.828427125\ldots$$

This approximation of π as 2.8284 is dreadfully inaccurate, worse even than the bicycle-tire estimate above. If we cannot do better than this, we should go back to the drawing board, or perhaps the sand tray.

But in the spirit of Archimedes we improve upon the first estimate by doubling the number of sides of the polygon, thereby getting a regular inscribed *octagon* and letting its perimeter be our next estimate of the circle's circumference. We double again to get a regular inscribed 16-gon, then a 32-gon, and so on. It is clear that at each step our estimate will be sharper. It is also clear that the major obstacle in our way is determining how the perimeter of one of these polygons relates to that of the next.

This obstacle is readily overcome with another application of the Pythagorean theorem. Figure C.4 shows a portion of a circle with center O and radius $r = 1$. The line segment AB of length a is a side of a regular inscribed n-gon. By bisecting AB at D and drawing a radius through D out to the circle at C, we generate segment AC, a side of a regular inscribed $2n$-gon. If b is the length of AC, we wish to determine the relationship between a and b; that is, between the side of a regular inscribed polygon and that of a regular inscribed polygon with twice as many sides.

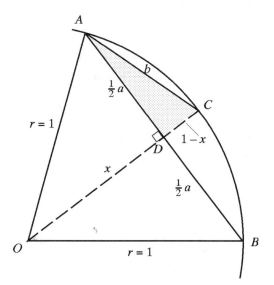

Figure C.4

Notice first that $\triangle ADO$ is a right triangle having hypotenuse of length $r = 1$ and leg AD of length $(1/2)a$. If x represents the length of leg OD, the Pythagorean theorem guarantees that

$$1^2 = \left(\frac{1}{2}a\right)^2 + x^2 = \frac{a^2}{4} + x^2 \rightarrow x^2 = 1 - \frac{a^2}{4} \rightarrow x = \sqrt{1 - \frac{a^2}{4}}$$

Because the length of CD is clearly the difference between the length of OC (a radius) and that of OD, we conclude that CD has length

$$1 - x = 1 - \sqrt{1 - \frac{a^2}{4}}$$

Applying the Pythagorean theorem once more to the shaded right triangle ADC yields

$$b^2 = \left(\frac{1}{2}a\right)^2 + (1-x)^2 = \frac{a^2}{4} + 1 - 2x + x^2$$

$$= \frac{a^2}{4} + 1 - 2\sqrt{1 - \frac{a^2}{4}} + 1 - \frac{a^2}{4} = 2 - 2\sqrt{1 - \frac{a^2}{4}}$$

because the $a^2/4$ terms cancel. We simplify this expression by moving the 2 from outside to inside the radical, squaring it in the process, to get

$$b^2 = 2 - 2\sqrt{1 - \frac{a^2}{4}} = 2 - \sqrt{4\left(1 - \frac{a^2}{4}\right)} = 2 - \sqrt{4 - a^2}$$

and finally arrive at

$$b = \sqrt{2 - \sqrt{4 - a^2}}$$

Now we return to the problem of estimating π. Recall that the side of our inscribed square was $s = \sqrt{2}$. This plays the role of a as we apply the formula above to calculate the side of a regular inscribed octagon:

$$b = \sqrt{2 - \sqrt{4 - a^2}} = \sqrt{2 - \sqrt{4 - (\sqrt{2})^2}} = \sqrt{2 - \sqrt{4 - 2}} = \sqrt{2 - \sqrt{2}}$$

Thus the *perimeter* of the octagon is $8 \times b = 8\sqrt{2 - \sqrt{2}}$ and we estimate π by

$$\pi = \frac{\text{circumference}}{\text{diameter}} \approx \frac{\text{perimeter}}{\text{diameter}} = \frac{8\sqrt{2 - \sqrt{2}}}{2} = 4\sqrt{2 - \sqrt{2}} = 3.061467459\ldots$$

Next we move to the 16-gon. This time $a = \sqrt{2 - \sqrt{2}}$, the side of the octagon just determined, and we use it to find b, the side of the 16-gon:

$$b = \sqrt{2 - \sqrt{4 - a^2}} \qquad = \sqrt{2 - \sqrt{4 - (\sqrt{2 - \sqrt{2}})^2}}$$

$$= \sqrt{2 - \sqrt{4 - (2 - \sqrt{2})}} \quad = \sqrt{2 - \sqrt{2 + \sqrt{2}}}$$

So the 16-gon's perimeter is

$$16 \times b = 16\sqrt{2 - \sqrt{2 + \sqrt{2}}}$$

and our improved estimate of π is

$$\pi = \frac{C}{D} \approx \frac{\text{perimeter}}{\text{diameter}} = \frac{16\sqrt{2 - \sqrt{2 + \sqrt{2}}}}{2} = 8\sqrt{2 - \sqrt{2 + \sqrt{2}}} = 3.121445153\ldots$$

Now we are getting somewhere. Another doubling and application of the formula yields the perimeter of a regular inscribed 32-gon as

$$32\sqrt{2 - \sqrt{2 + \sqrt{2 + \sqrt{2}}}}$$

so that

$$\pi \approx \frac{\text{perimeter}}{\text{diameter}} = 16\sqrt{2 - \sqrt{2 + \sqrt{2 + \sqrt{2}}}} = 3.136548491\ldots$$

And on we go. Obviously we can repeat the procedure as often as we wish. In fact, the developing pattern makes the transition from one step to the next quite painless.

With the assistance of a hand calculator, we doubled seven more times, to a 64-gon, a 128-gon, a 256-gon, a 512-gon, a 1024-gon, a 2048-gon, and a 4096-gon. It is evident that a regular 4096-gon, though not coinciding with the circle in which it is inscribed, is pretty close. This yields an estimate of π as follows:

$$\pi = \frac{C}{D} \approx \frac{\text{perimeter}}{\text{diameter}} =$$

$$= 2048\sqrt{2 - \sqrt{2 + \sqrt{2 + \sqrt{2 + \sqrt{2 + \sqrt{2 + \sqrt{2 + \sqrt{2 + \sqrt{2 + \sqrt{2}}}}}}}}}}$$

$$= 3.141594618\ldots$$

This expression, whose very appearance suggests a certain mathematical artistry, is accurate to five decimal places. More important is that we know how to get a *more* accurate estimate: Just carry the pattern a step further or, should the mood strike, 50 steps further. In this manner the constant π can be approximated as closely as we wish.

This basic approach using regular polygons dates back 22 centuries to Archimedes. But it has one liability: the need to calculate square roots of square roots of square roots. With each additional doubling of the number of sides, we embed another square root in the design and complicate the process accordingly. Archimedes, who had neither a decimal system nor a hand calculator, had to weather the storm of square roots by finding fractions of roughly equal value. He stopped with the 96-gon. That he was able to get *that* far is a testimony to his genius.

But is there an easier, more efficient route to the same end? The answer is yes, although it remained obscure until the discovery of calculus and infinite series during the seventeenth century. Only then could mathematicians find truly efficient approximations of π. Although this topic is one of considerable subtlety, we would like to give at least a sense of the line of attack.

There is an important function called the inverse tangent (denoted by $\tan^{-1}x$) whose origins lie in the domain of trigonometry and need not concern us here. What *is* important is that we can express $\tan^{-1}x$ as the infinite series

$$\tan^{-1}x = x - \frac{x^3}{3} + \frac{x^5}{5} - \frac{x^7}{7} + \frac{x^9}{9} - \frac{x^{11}}{11} + \frac{x^{13}}{13} - \cdots$$

where the summation process continues indefinitely in the obvious pattern. The further we carry the arithmetic, the closer we get to the true value of $\tan^{-1}x$.

But what does this have to do with π? It can be proved, using trigonometry, that

$$\pi = 4\left[\tan^{-1}\frac{1}{2} + \tan^{-1}\frac{1}{5} + \tan^{-1}\frac{1}{8}\right]$$

Then we approximate $\tan^{-1}1/2$, $\tan^{-1}1/5$, and $\tan^{-1}1/8$ by putting, respectively, $x = 1/2, x = 1/5$, and $x = 1/8$ into the series shown above. Carrying the computations in each series through seven terms yields

$$\pi = 4\left[\tan^{-1}\frac{1}{2} + \tan^{-1}\frac{1}{5} + \tan^{-1}\frac{1}{8}\right]$$

$$\approx 4\left[\left(1/2 - \frac{(1/2)^3}{3} + \frac{(1/2)^5}{5} - \frac{(1/2)^7}{7} + \frac{(1/2)^9}{9} - \frac{(1/2)^{11}}{11} + \frac{(1/2)^{13}}{13}\right)\right.$$

$$+ \left(1/5 - \frac{(1/5)^3}{3} + \frac{(1/5)^5}{5} - \frac{(1/5)^7}{7} + \frac{(1/5)^9}{9} - \frac{(1/5)^{11}}{11} + \frac{(1/5)^{13}}{13}\right)$$

$$+ \left.\left(1/8 - \frac{(1/8)^3}{3} + \frac{(1/8)^5}{5} - \frac{(1/8)^7}{7} + \frac{(1/8)^9}{9} - \frac{(1/8)^{11}}{11} + \frac{(1/8)^{13}}{13}\right)\right]$$

$$= 4(0.785399829\ldots) = 3.141599318\ldots$$

Like our previous estimate, this is accurate to a good number of decimal places. But whereas the previous approximation introduced many square roots—each of which required its own approximation procedure—this estimate is computed without a square root anywhere in sight! By introducing the infinite series for $\tan^{-1}x$, mathematicians avoided the square root nightmare altogether.

This discovery, made about three centuries ago, allowed for tremendous progress in the calculation of π. But a more recent advance must also be noted: the use of computers to facilitate the calculations. In 1948 (before computers), π was known to 808 decimal places. A year later, the ENIAC computer—abysmally primitive by modern standards—pushed the accuracy to 2037 places.[2] The magnitude of the improvement suggests a truism: All subsequent computations of π would be done using computers. Indeed, digit searching has become a kind of passion for a small but dedicated band of number crunchers. In rapid succession, accuracy grew to 100,000 digits, a million digits, and then a stunning billion digits. Such calculations tend to be performed on powerful supercomputers at major universities or large research centers.

Bucking this trend, however, are David and Gregory Chudnovsky, two brilliant and eccentric brothers who computed π to two billion digits by wiring together mail-order computer components in their Manhattan apartment. Their project filled tabletops with computer equipment and hallways with cables, and the heat generated by all this electronic gadgetry raised the apartment temperature to hellish proportions. Nonetheless, the Chudnovskys pulled it off. The contrast between their approach and that of the universities is the mathematical counterpart of David (and Gregory) versus Goliath, although in this case the underdogs are armed with silicon slingshots.[3]

If the Chudnovskys of New York are lone wolves who successfully attacked π, Dr. E. J. Goodwin's solitary assault was spectacularly unsuccessful. His story is familiar to mathematicians but always bears a good retelling.

It occurred during the last years of the nineteenth century. Dr. Goodwin lived in Solitude, Indiana, a well-named town that was tiny, remote, and dull. To occupy his time, the good doctor dabbled in mathematics, unfortunately with more enthusiasm than ability. He believed he had made a tremendous discovery about the relationship between a circle's area and its circumference, which in turn would imply a tremendous discovery about π itself.

Great mathematical advances should be shared with the academic community, but Dr. Goodwin adopted a different tactic. He took his case into the political, not the scholarly, arena by asking his representative in the Indiana General Assembly to introduce the following as House Bill 246 of the year 1897: "Be it enacted by the General Assembly of the state of Indiana, That it has been found that a circular area is equal to the square on a line equal to the quadrant of the circumference."[4] Apparently, the political leaders of 1897 were no more adept mathematically than their modern counterparts, for they found this perfectly acceptable. But what did it mean?

As shown in Figure C.5, Goodwin's bill asserted that the area of the circle on the left equalled that of the square on the right, where the latter has each side exactly as long as the circle's quadrant, that is, as one-fourth of its circumference. If we denote

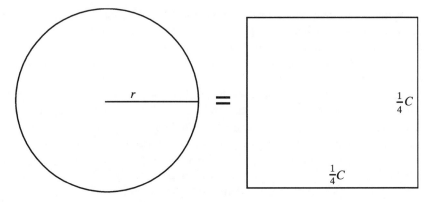

Figure C.5

the circle's radius by r and circumference by $C = 2\pi r$, we know that the area of the circle is πr^2 and that of the square is

$$\left(\frac{1}{4} C\right)^2 = \left(\frac{1}{4} 2\pi r\right)^2 = \left(\frac{1}{2} \pi r\right)^2 = \frac{1}{4} \pi^2 r^2$$

If, as Goodwin claimed, these areas are equal, then it follows that

$$\pi r^2 = \text{area(circle)} = \text{area(square)} = \frac{1}{4} \pi^2 r^2$$

Upon cross-multiplying, we get $4\pi r^2 = \pi^2 r^2$ and cancellation of πr^2 from each side yields the remarkable result that $\pi = 4$.

Although Archimedes would have turned over in his grave, none of the Hoosier lawmakers was much troubled by this conclusion. To them, the language probably sounded abstruse enough to be unassailable. The bill passed its first committee, strangely but perhaps appropriately the House Committee on Swamp Lands. On February 2, 1897, it was passed by the House Committee on Education. Three days later, the entire Indiana House of Representatives voted "aye" on Goodwin's assertion that $\pi = 4$.

At some point along the way, the matter began to attract attention from the press. The *Indianapolis Sentinel* weighed in with its support:

> The bill . . . is not intended to be a hoax. Dr. Goodwin . . . and State Superintendent of Public Instruction Geeting believe that it is the long-sought solution. . . . Dr. Goodwin, the author, is a mathematician of note. He has it copyrighted and his proposition is that if the legislature will endorse the solution, he will allow the state to use [it] free of charge.[5]

Besides revealing that the superintendent of public instruction supported the bill, this passage gives a possible rationale for these strange goings-on: The legislators coveted the national or even international royalties owed to Indiana by all who would use this new value of π.

House Bill 246 moved to the Senate's Committee on Temperance. This it passed on February 12 and was now on its way to full Senate approval and the status of law.

Fortunately the bill was derailed at the last moment. Its downfall was largely due to the efforts of Purdue mathematician C. A. Waldo, who was in Indianapolis at the time. Waldo gave this third-person recollection of what happened when he visited the capitol: "A member then showed the writer a copy of the bill . . . and asked him if he would like an introduction to the learned doctor, its author. He declined the courtesy with thanks, remarking that he was acquainted with as many crazy people as he cared to know." [6]

With the professor's negative appraisal, support for the legislation collapsed. On the afternoon of February 12, the Senate indefinitely postponed the proposal, and it remained legal for π to equal 3.14159 . . . in Indiana as elsewhere. Senator Hubbell, an insightful opponent of the bill, summed it up nicely when he grumbled, "The Senate might as well try to legislate water to run up hill."

Circles and π have thus intrigued people from the sand trays of Archimedes to the legislative halls of Indianapolis. We shall meet both concepts again in later pages of this book, for they are central to the mathematical enterprise. For now, we leave the reader with the first 30 decimal places of one of the world's great numbers:

$$\pi = 3.141592653589793238462643383279 \ldots$$

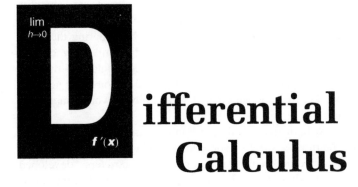

Differential Calculus

During the year 1684 a mathematical paper appeared in an issue of the journal *Acta Eruditorum*. Its author was Gottfried Wilhelm Leibniz, a German scholar and diplomat with broad interests and virtually limitless abilities. The paper was a densely packed mixture of Latin words and mathematical symbols, and contemporary readers probably found little in it that was comprehensible. From the modern perspective, the best clue to the subject matter is a word appearing inobtrusively near the end of the title: *calculi*.

This was the first published version of calculus. Its title translates as "A New Method for Maxima and Minima, as well as Tangents, which is impeded neither by Fractional nor Irrational Quantities, and a Remarkable Type of Calculus for This."[1] Here the word *calculus* means "set of rules," in this case rules applicable to problems about maximum and minimum values and tangents that Leibniz proclaimed to be impervious to the onslaughts of fractional or irrational quantities. Because of the significance of his discovery, the word *calculus* has achieved mathematical immortality. In truth, when mathematicians want to lend special emphasis to the subject, they call it "*the* calculus," which sounds even more awesome.

Awesome it is. In the traditional undergraduate curriculum, calculus serves as an entry (and, unfortunately for some, as a barrier) to higher mathematics. It has become an indispensable tool for engineers, physicists, chemists, economists, and many others. Calculus was certainly the crowning achievement of seventeenth-century mathematics, and many regard it as the crowning achievement of mathematics, period. John von Neumann (1903–1957), one of the twentieth century's most influential mathematicians, wrote that "The calculus was the first achievement of modern mathematics, and it is difficult to overestimate its importance."[2] (Note von Neumann's reference to *the* calculus.)

Leibniz's 1684 paper was about *differential* calculus, one of the two branches of the subject. The other, *integral* calculus, was introduced by Leibniz in a 1686 issue of the same periodical and is our subject in Chapter L.

Before taking a look at differential calculus, we should say a brief word about its origins. Although it was Leibniz who first published an account of calculus in the mid-1680s, it was Isaac Newton who first developed the subject during the years 1664–1666. Newton, a student at Trinity College of Cambridge University, created what he called "fluxions," a body of rules with which he, too, could find maxima and minima, as well as tangents, and which were not impeded by fractional nor irrational quantities. In short, his fluxions anticipated what Leibniz would publish as calculus two decades later.

Modern scholars accord both individuals the honor of independent discovery. But mathematicians of the time, suspecting some kind of plagiarism at work, were far less generous about assigning credit. An acrimonious controversy developed between the British, staunch advocates of Newton's priority claims, and the continental mathematicians, equally staunch defenders of Leibniz. This dispute, one of the most unfortunate episodes from the entire history of mathematics, is considered at length in Chapter K.

What was it Newton and Leibniz discovered? At the heart of differential calculus are the notions of slope, a concept regularly introduced in high school algebra, and tangent, a key idea from high school geometry. The latter appeared in Leibniz's title, but it is with the former that our treatment begins.

Suppose we have a straight line graphed in the coordinate plane. We might investigate the x-coordinates or the y-coordinates separately, but it is often beneficial to examine how x and y vary jointly. For instance, if x increases by 4 units, what happens to the associated y values?

A little thought indicates that the answer depends on the steepness of the line in question. In Figure D.1, the left-hand line rises gradually, so an increase of 4 units in x (that is, 4 units horizontally) results in very little change in y (that is, very little vertical change). But for the steeper right-hand graph, an increase of 4 in x yields a significant rise in y.

To put this concept on firmer mathematical ground, we define the *slope* of a line by:

$$\text{slope} = \frac{\text{change in } y}{\text{change in } x} = \frac{\text{rise}}{\text{run}}$$

If a straight line has slope $= 2/5$, then as x grows by 5 units, y simultaneously grows by 2 units, a fairly gradual rise. On the other hand, a slope of $5/2$ means as x increases by 2, y increases by a hefty 5 units, a much steeper climb. If we were required to haul a piano up an inclined ramp, we would prefer a slope of $2/5$ to one of $5/2$.

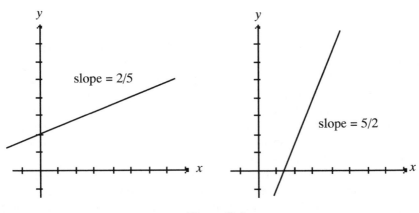

Figure D.1

The usual symbol for slope is *m*, and as illustrated in Figure D.2, the formal definition for the slope of a line passing through the points (x_1, y_1) and (x_2, y_2) is

$$m = \frac{\text{change in } y}{\text{change in } x} = \frac{y_2 - y_1}{x_2 - x_1}$$

For a line with slope 5/2, an increase of 2 units horizontally results in a rise of 5 vertically. Hence if *x* were to grow by $3 \times 2 = 6$ units, the accompanying growth in *y* would be $3 \times 5 = 15$ units. Likewise—and this is critical for interpreting slope—a rightward increase in *x* of a *single* unit would lead to an increase in *y* of $5/2 = 2.5$ units. For a line with slope 2/5, an increase of 1 unit in *x* results in an increase of $2/5 = 0.4$ units in *y*. Thus, we can think of the slope of a line as the change in *y* per unit change in *x*. Quite simply, the slope tells us what *y* does when *x* increases by 1.

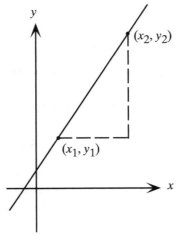

Figure D.2

All of this might seem empty of any real-world significance, but nothing could be further from the truth. Suppose, for instance, we are considering the motion of an airplane, where x represents the amount of time the plane has been aloft and y is the distance it has flown during those x hours. Assuming that the graph of this x-y relationship is a straight line, we interpret the line's slope to be the change in distance (change in y) per unit change in time (change in x). That is, the slope represents the plane's *velocity* measured in miles per hour. That velocity is a subject of importance to airplane pilots is undeniable. That it is intimately linked to the abstract mathematical concept of slope gives an indication of why this idea holds such importance beyond the realm of pure mathematics.

Or consider an economic problem. We are plotting two variables related to a manufacturing process: x is the number of units produced and y is the profit resulting from the sale of x units. If the x-y relationship is linear, we interpret the slope as the change in profit per unit change in sales; that is, the additional profit we make from each additional item sold. Economists are so enamored of this concept that they give it a special name, *marginal profit,* and its value can determine the course of great industries.

There are many other examples of slopes arising in natural ways. Such measures as miles per gallon, feet per second, or price per pound indicate that a slope is lurking nearby. Without doubt, some of the most important applications of mathematics involve the rate of change of one quantity with respect to another that is embodied in the idea of slope.

For our examples thus far, a unit increase in x resulted in an associated *increase* in y. Graphically, this means the line is climbing as we move toward the right. But not all linear relationships are of this type. We obviously can encounter cases in which an increase in x leads to a corresponding decrease in y. Returning to the airplane example, we might let x be the time aloft and y be the distance of the plane from its destination. In this case, as x grows, y shrinks. Such situations are illustrated by the line on the left in Figure D.3, where, as x increases by 2, y decreases by 5. Here

$$m = \frac{\text{change in } y}{\text{change in } x} = \frac{-5}{2} = -2.5$$

A final case, and one of particular importance for differential calculus, is that of the horizontal line on the right in Figure D.3. Here, an increase in x results in neither an increase nor a decrease in y, because y is unchanging. Consequently,

$$m = \frac{\text{change in } y}{\text{change in } x} = \frac{0}{\text{change in } x} = 0$$

In summary, an uphill line has positive slope; a downhill line has negative slope; and a horizontal line, at the boundary between uphill and downhill, has slope of zero, at the boundary between positive and negative. It is all very consistent.

Unfortunately this theory is applicable only to straight lines, which exhibit the same inclination—the same slope—throughout their length. Lines are certainly of

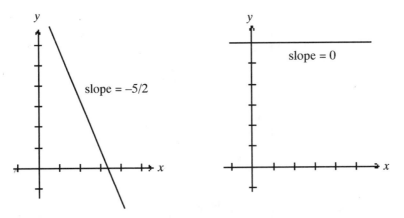

Figure D.3

great importance in mathematics, but it is obvious that many real-world phenomena produce variable, nonlinear behavior. Airplanes do not fly at a constant velocity; manufacturing processes do not exhibit a constant marginal profit. How, if at all, can we determine slopes of *curves*? In addressing this question, we finally move into the realm of differential calculus.

To illustrate the problem, consider the graph of the parabola $y = x^2 - 4x + 7$ shown in Figure D.4. When $x = 3$, we find $y = 3^2 - 4(3) + 7 = 4$ and label the point $(3, 4)$ on the curve as A.

Clearly a parabola does not have a constant slope throughout. As we move along the curve, we constantly change direction, from a downhill plunge on the left, to a leveling out at the bottom, to an ever steeper rise toward the right. A basic principle is evident: For curves, unlike lines, the slope changes from point to point.

So how can we determine the slope of this curve at A? Geometrically, it seems reasonable to draw the *tangent line* to the parabola at A and interpret the slope of the (curving) parabola to be the slope of the (linear) tangent at this point. The reasonableness of this approach is suggested by the following scenario.

Suppose we are riding in a tiny vehicle along this parabolic path. We have descended from the left, leveled out, and now are climbing toward the right, all the time heading more steeply upward. Precisely as we reach the point $(3, 4)$, we suddenly fly out of the vehicle and continue in a straight line (with the arrow in Figure D.4) as the vehicle curves up and away on its parabolic path. Our line of flight is thus tangent to the curve at $(3, 4)$, and the *slope* of this tangent line is what we shall mean by the slope of the parabola at the point A.

This much is easy. Less obvious is how to find the tangent's slope. Before examining the solution, we should mention where the difficulty lies. Because slope is defined by

$$m = \frac{y_2 - y_1}{x_2 - x_1}$$

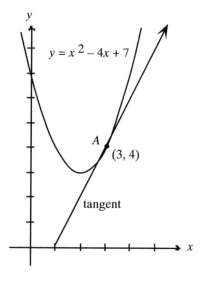

Figure D.4

two points on a line are necessary in order to compute it. In the example above, however, we know only a single point on the tangent line, namely $A = (3, 4)$ itself. If we knew the identity of just one other point on the tangent, finding its slope would be a snap. Without such knowledge it may seem as though we have reached a dead end, but differential calculus provides a way around this roadblock by approaching the slope of the tangent line indirectly. It is a wonderfully ingenious line of attack.

For our problem, where we want the slope of the tangent line at $x = 3$, we first consider what happens at $x = 4$. At the moment there is no way to know the point on the *tangent line* corresponding to $x = 4$, but we can determine the point on the *parabola* for $x = 4$, namely $y = 4^2 - 4(4) + 7 = 7$. We label the point $(4, 7)$ as B in Figure D.5, which shows an enlargement of the critical portion of the curve. It is easy to find the slope of the straight line, called the *secant,* connecting A and B:

$$m = \frac{y_2 - y_1}{x_2 - x_1} = \frac{7 - 4}{4 - 3} = \frac{3}{1} = 3$$

This is a perfectly simple calculation; unfortunately it is not the slope of the tangent line itself but that of the secant line, a rough approximation. How can we improve the estimate?

Why not choose a point on the parabola closer to A than was B? Perhaps let $x = 3.5$. The associated y value is $3.5^2 - 4(3.5) + 7 = 5.25$, so the point C with coordinates $(3.5, 5.25)$ is on the parabola. The secant line connecting A and C has slope

$$m = \frac{y_2 - y_1}{x_2 - x_1} = \frac{5.25 - 4}{3.5 - 3} = \frac{1.25}{0.5} = 2.50$$

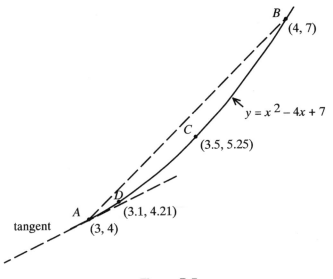

Figure D.5

If one imagines drawing the line between A and C on Figure D.5, it will clearly be a better approximator to the tangent than was our first try, that between A and B. Consequently, the slope of 2.50 is closer to the slope of the tangent than was our first estimate of 3.00.

The next step should be predictable: Take a point on the parabola even closer to A. For instance, let $x = 3.10$ so that $y = 3.10^2 - 4(3.10) + 7 = 4.21$, and let D be the point $(3.10, 4.21)$. The secant line connecting A and D lies quite close to the desired tangent, and the slope of the former is

$$m = \frac{y_2 - y_1}{x_2 - x_1} = \frac{4.21 - 4}{3.10 - 3} = \frac{0.21}{0.10} = 2.10$$

We continue in this fashion, letting our points slide down along the parabola toward A and computing the slopes of the associated secant lines as we go. Such a string of computations appears in the accompanying chart.

Point on Parabola	Slope of Secant Connecting This Point with A
(4.0, 7)	3.0
(3.5, 5.25)	2.5
(3.10, 4.21)	2.10
(3.01, 4.0201)	2.01
(3.0001, 4.00020001)	2.0001
.	.
.	.
.	.

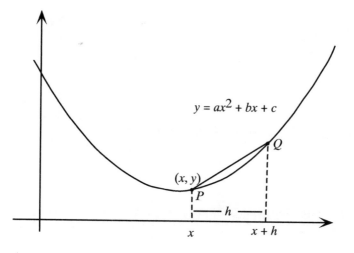

$$y = ax^2 + bx + c$$

Figure D.6

A pattern is evident. As our point moves along the parabola toward $A = (3, 4)$, the associated secant lines are rotating toward ever better approximations of the tangent line, and their slopes provide ever more accurate estimates of the elusive slope of the tangent. In our example, we readily surmise that the *exact* slope of the tangent in question is the number toward which these estimates are tending: The slope of the tangent to the parabola $y = x^2 - 4x + 7$ at the point $(3, 4)$ appears to be 2.

So far, so good. But what if we needed the slope of the tangent to the same parabola at the point $(1, 4)$? We would have to run through similar calculations and prepare a similar chart. And what if we were then given a dozen additional points at which the slope of the tangent was required? We would face a dozen more charts, and the whole business would become excessively tedious. Can the process of finding slopes of tangents be streamlined?

Of course. This, in fact, is what was accomplished by the rules Leibniz described in his 1684 article. The streamlining requires us to adopt a slightly more general—which is to say, more algebraic—viewpoint. Instead of focusing on the point $(3, 4)$, we want to develop a formula for the slope of the tangent at *any* point P on any parabola $y = ax^2 + bx + c$.

Let P have coordinates (x, y), where $y = ax^2 + bx + c$. As above, choose a point near (x, y) and use the slope of the secant to approximate the slope of the tangent.

As Figure D.6 indicates, it is customary to denote the first coordinate of the "nearby" point as $x + h$. In so doing, we regard h as a tiny but unspecified amount, a small increment that moves us just a bit beyond x. The associated point on the parabola is labeled Q in the diagram. To find its second coordinate, we merely substitute $x + h$ into the parabola's equation; in other words, replace each occurrence of x by $x + h$. This gives Q's second coordinate as

$$a(x + h)^2 + b(x + h) + c = a(x^2 + 2xh + h^2) + b(x + h) + c$$
$$= ax^2 + 2axh + ah^2 + bx + bh + c$$

so Q is the point $(x + h, ax^2 + 2axh + ah^2 + bx + bh + c)$. The reader will notice that the algebraic intensity of the problem has been turned up a notch or two, but the benefits of a general formula will be worth the effort.

The next step is to use the formula for m to determine the slope of the secant line between P and Q:

$$m = \frac{y_2 - y_1}{x_2 - x_1} = \frac{(ax^2 + 2axh + ah^2 + bx + bh + c) - (ax^2 + bx + c)}{(x + h) - x}$$

$$= \frac{ax^2 + 2axh + ah^2 + bx + bh + c - ax^2 - bx - c}{x + h - x}$$

$$= \frac{2axh + ah^2 + bh}{h}$$

after combining like terms in numerator and denominator

$$= \frac{h(2ax + ah + b)}{h}$$

after factoring h out of the numerator

$$= 2ax + ah + b$$

upon cancelling the common h

In short, for any small increment h, the slope of the secant line between P and Q is given algebraically by $2ax + ah + b$. But the idea of "sliding" Q toward P along the parabola amounts to nothing more than letting h tend ever closer toward zero. In other words, we determine the *exact* slope of the tangent by taking the limit of the slopes of the secant as h approaches zero. For our example, the slope of the tangent is therefore given by:

$$\lim_{h \to 0} (2ax + ah + b) = 2ax + a(0) + b = 2ax + b$$

because a, b, and x remain unchanged as h moves toward zero. (The symbol $\lim_{h \to 0}$ is read "the limit as h approaches 0.")

We note in passing that we can apply this general formula to the specific parabola $y = x^2 - 4x + 7$ cited above. Here, $a = 1$, $b = -4$, and $c = 7$. Thus, at the point A where $x = 3$, the slope of the tangent line will be $2ax + b = 2(1)(3) + (-4) = 2$, as our chart suggested. If we wanted the slope of the tangent at $(1,4)$, we merely let $x = 1$ and get a slope of $2(1)(1) - 4 = -2$; the graph confirms the parabola is descending at this point, consistent with the negative slope.

To repeat: The slope of the tangent to a curve is taken to be the *limit* of the slopes of the associated secants as h goes to zero. This limit is called the *derivative;* the process of finding the derivative is called *differentiation;* and the branch of mathematics investigating these things is called *differential calculus.*

One goal of differential calculus is to develop ever more general formulas. We certainly do not want to be restricted to treating parabolas. Following a procedure analogous to that above, mathematicians start with an unspecified function $y = f(x)$ and seek the slope of its tangent at any point (x, y) upon it. As before, we choose a nearby point on the curve whose first coordinate is $x + h$ and whose second coordinate is consequently $f(x + h)$; next, determine the slope of the secant line:

$$m = \frac{y_2 - y_1}{x_2 - x_1} = \frac{f(x + h) - f(x)}{(x + h) - x} = \frac{f(x + h) - (x)}{h}$$

and finally take the limit of this quotient as $h \to 0$.

Leibniz denoted the derivative by dy/dx. A competing notation was later introduced by Joseph Louis Lagrange (1736–1813), who used the symbol $f'(x)$ for the derivative of $f(x)$. With this, we arrive at the basic formula found in all books on differential calculus:

$$f'(x) = \lim_{h \to 0} \frac{f(x + h) - f(x)}{h}$$

Starting with this general definition, we can find derivatives of a vast array of functions. When differentiating a power of x—that is, a function of the form x^n—a beautiful pattern emerges, namely:

$$\text{if } f(x) = x^n, \text{ then } f'(x) = n\, x^{n-1}$$

In words, this says that finding the derivative of a power x^n requires nothing more than moving the exponent down in front as a coefficient and lowering the power by one. Thus the derivative of x^5 is $5x^4$ and that of x^{19} is $19x^{18}$. This is certainly a strange and wonderful rule. Somehow, buried deep within the mathematics, properties of curves and their tangents translate into something so simple.

At this point it is proper to issue a few warnings about the definition of the derivative. First, although derivatives of some functions emerge easily from the associated algebra, there are plenty of functions for which the derivative formula leads into a tangled mathematical mess. Worse, for some functions the derivative fails even to exist at one or more points. For such functions, it is impossible to assign *any* number as the slope of the tangent to the curve at the point in question.

An example is illustrated in Figure D.7. At the point (2,1) this graph has a sharp corner. There is no way to draw a unique tangent to the curve at (2,1) because it abruptly changes direction there. But if we cannot draw a tangent, we certainly cannot determine the tangent's slope, and this is what is meant by its derivative. This function, indeed any function with a jagged graph, has no derivative at its corners.

Our example indicates that subtle and difficult issues can arise in connection with the derivative. These usually involve the concept of "limit," an idea with which mathematicians have grappled in one form or another since classical times. The theoretical meaning of limit, upon which the definition of the derivative rests, is by no means trivial. Needless to say, we shall not pursue the concept into these logical

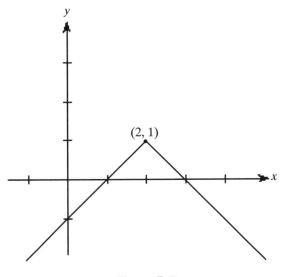

Figure D.7

depths. Neither, we add, did Leibniz. He was happy to exploit the more immediate benefits of his "New Method of Maxima and Minima, as well as Tangents" without worrying too much about their theoretical underpinnings.

We have spent a good deal of time on tangents. We conclude this chapter with an application of differential calculus to maxima and minima.

At the outset, we stress that knowing how big or how small a function can get— in other words, knowing its maximum or minimum values—can be of extreme importance to both the theoretician and the user of mathematics. Under what conditions can we maximize profit, or minimize gasoline consumption? Questions of extremes lie at the heart of decisions made daily in the world around us. That differential calculus provides a tool for answering these questions suggests something of its power.

To see how this works, consider the graph of a generic function $y = f(x)$ in Figure D.8. This example is certainly not linear, for it rises and falls as x moves rightward, but two of its points deserve special notice. These are M, where the curve reaches a maximum, and N, where it attains a minimum. It would certainly be of interest to determine the coordinates of M and N.

But how is this done? The key to finding maxima and minima lies in our prior discussion of slope: At the top of a crest or the bottom of a valley, the tangent to the curve will be *horizontal*, and a horizontal line, as we noted, has a slope of zero. Therefore our search for a maximum or minimum leads precisely to those points at which the slope of the tangent line—in other words, the derivative—equals zero. Stated algebraically, our task is to solve the equation $f'(x) = 0$ and we should be well on our way to finding a function's extreme values.

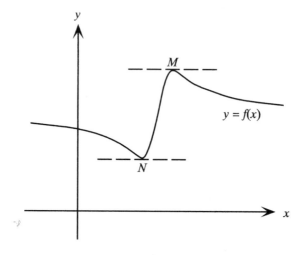

Figure D.8

As an illustration, we examine an assertion of the Italian mathematician Gerolamo Cardano (1501–1576) to which we shall return in a different context in Chapter Z. While considering a problem in algebra, Cardano claimed that there do not exist two real numbers summing to 10 and multiplying to 40. With differential calculus, we can easily verify his claim.

That is, let one of the two real numbers be x and the other be z. The fact that their sum is 10 translates into the equation $x + z = 10$, from which it immediately follows that $z = 10 - x$. We wish to determine how large the product xz can become. Clearly, $xz = x(10 - x)$, so we introduce the product function

$$f(x) = xz = x(10 - x) = -x^2 + 10x$$

and apply calculus to find its maximum.

We already showed the derivative of the general quadratic $f(x) = ax^2 + bx + c$ is $f'(x) = 2ax + b$; hence, the function $f(x) = -x^2 + 10x$ has derivative $f'(x) = -2x + 10$ (because $a = -1$, $b = 10$, and $c = 0$). To find the maximum product, we need only locate those values of x where the curve has a horizontal tangent. We thus set $f'(x) = 0$ and solve the resulting equation for x:

$$0 = \text{slope of tangent} = f'(x) = -2x + 10 \rightarrow 2x = 10 \text{ and so } x = 5$$

The graph of the product function $f(x) = -x^2 + 10x$ in Figure D.9 supports this conclusion, for the parabola tops out when $x = 5$. At this point the product of x and z is $f(x) = xz = 5(10 - 5) = 25$, and this is as big as the product can get. In other words, two real numbers summing to 10 have a *maximum* possible product of 25. Cardano was correct in saying that no such numbers could possibly multiply to 40.

The previous examples have given a bit of the flavor, but have barely scratched the surface, of differential calculus. The subject has applications to a bewildering

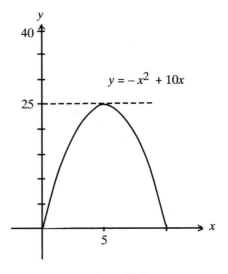

Figure D.9

array of problems, and it should come as no surprise that we meet it again in later chapters. But for now we must leave the topic, a concept of enormous mathematical importance and one that has justified Leibniz's enthusiastic description of "A Remarkable Type of Calculus" from three centuries ago.

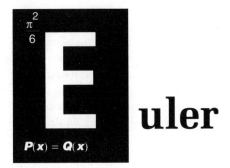

Euler

In previous chapters we have met some mathematical superstars: Euclid, the Bernoullis, Archimedes. In this spirit we devote the present chapter to one of the greatest mathematicians in history, Leonhard Euler (whose name rhymes with *boiler* not *ruler*). Astonishingly prolific, Euler created a body of mathematics whose sheer volume defies belief. But what endears him to his successors is not so much the mass of his writings as their richness, beauty, and penetrating insight.

A short chapter cannot even begin to do justice to Euler's legacy. Where a modern mathematician can build a respectable reputation with a dozen (or fewer) publications of moderate (or lesser) quality, Euler produced nearly 900 treatises, books, and papers. When the dust had settled, his output far surpassed in both quantity and quality that of scores of mathematicians working many lifetimes. It is estimated that he published an average of 800 pages of new mathematics per year over a career that spanned six decades.[1] In all of history, no mathematician has been able to think that fast; for that matter, most people can't even *write* that fast. Without question, Euler possessed a mental agility, a quickness, an ingenuity matched by only the smallest handful of mathematicians from antiquity down to the present day. Like Michelangelo or Einstein, he was an undisputed master.

Starting in 1911 scholars began to publish Euler's collected works under the title *Opera Omnia*. This was one of the most ambitious publication projects ever begun. To date there are well over 70 volumes on the shelves (but who's counting?) and new ones will appear sporadically well into the twenty-first century. As a typical volume runs to 500 large pages and weighs about 4 pounds, the overall bulk of his *Opera Omnia* stands at more than 300 pounds! No other mathematician can match this poundage.

So, Euler was prolific. In addition he was extremely broad in his interests. Accomplished in such established subjects as number theory, calculus, algebra, and geometry, he also gave birth almost single-handedly to new branches of mathematics such as graph theory, the calculus of variations, and combinatorial topology. He was influential in establishing the legitimacy of complex numbers, which we discuss in Chapter Z, and of elevating to prominence the idea of "function" that now unifies much of mathematics.

Euler excelled in applied mathematics as well. He attacked problems of mechanics, optics, electricity, and acoustics with his powerful mathematical weaponry and thereby explained natural phenomena ranging from the motion of the moon to the flow of heat to the underlying structure of music (although regarding the latter it was said that Euler's work contained "too much geometry for musicians and too much music for geometers.")[2] More than half the volumes of the *Opera Omnia* deal with applications.

It is important to observe that Euler was a skillful expositor whose choice of notation and terminology soon became the standard of the discipline. Consequently his mathematical writings "look" modern because all who followed have written mathematics as Euler did. Among his texts, the most revered is the *Introductio in analysin infinitorum* of 1748. This book, writes historian of mathematics Carl Boyer,

> is probably the most influential textbook of modern times. It is the work which made the function concept basic in mathematics. It popularized the definition of logarithms as exponents and the definitions of the trigonometric functions as ratios. It crystallized the distinction between algebraic and transcendental functions and between elementary and higher functions. It developed the use of polar coordinates and of the parametric representation of curves. Many of our commonplace notations are derived from it. In a word, the *Introductio* did for elementary analysis what the *Elements* of Euclid did for geometry.[3]

No less a mathematician than Gauss, describing his first exposure to Euler's works, recalled becoming "animated with fresh ardor," and "fortified in my resolution to push forward the boundaries of this wide department of science."[4]

The portrait of Euler shown here is well worth remembering. If anyone ever carves a Mount Rushmore of mathematics, Euler's likeness will occupy a prominent position.

Euler was born in Basel, Switzerland, in 1707. While still an adolescent he studied with Johann Bernoulli at a time when the latter was recognized as one of the world's great mathematicians. Unquestionably this was to Euler's advantage, even if he had to contend with the prickly Bernoulli personality. (One imagines the grumpy Johann, after a session with young Euler, muttering the eternal complaint of the teacher: "Students just aren't as good as they used to be.")

But Johann Bernoulli had no cause to complain, for few teachers have ever had a student like this. Euler finished college at the age of 15. Four years later, he made his first international splash by winning a prize from the Parisian Academy of Sciences. The academy had issued as a challenge the problem of determining the optimum

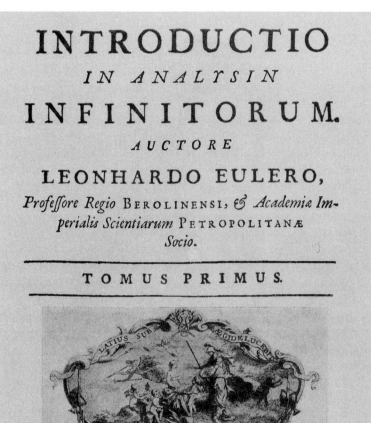

Frontispiece of Euler's *Introductio*
(Courtesy of Lehigh University Library)

placement of masts upon a sailing ship, and Euler's solution was deemed worthy of recognition. It is always observed that the Swiss Euler won his prize in spite of the fact that Switzerland is not known as a seafaring power. The power, in this case, was mathematical rather than nautical.

Leonhard Euler
(Courtesy of Lehigh University Library)

During 1727 Euler, barely 20, traveled to Russia to assume a chair in the recently founded St. Petersburg Academy. There he remained until 1741, when an offer from the Berlin Academy of Frederick the Great proved more attractive. For a quarter of a century Euler worked at the German academy, where he overlapped such notables as d'Alembert, Maupertuis, and Voltaire. Then in 1766 he returned to St. Petersburg for good. Euler died there in 1783, still scientifically active at the age of 76.

By all accounts Euler was a modest and unassuming individual, one who valued family life and easily made friends. His good nature persisted in spite of the painful loss of vision beginning in 1735 and progressing to nearly total blindness by 1771. Remarkably, these difficulties did not sap his spirit nor impede his research. He simply kept at it, even if it meant dictating to an amanuensis the formulas and equations he could see only in his mind's eye. The record of his mathematical discoveries shows that blindness was in no way a barrier to his productivity, and to this day his triumph in the face of adversity remains an enduring legacy.

It is absurd to try to summarize Euler's mathematical discoveries in a few pages. We instead describe his contributions to a few mathematical subdisciplines, leaving the reader to multiply these by many thousands to get some sense of the scope of his work.

We began with Chapter A on the higher arithmetic, so we first mention one of Euler's number theoretic contributions. This was a branch of mathematics to which Euler was not immediately drawn, but once he took the plunge he was hooked. Four volumes of his *Opera Omnia*, running to some 1,700 pages, contain Euler's number theoretic papers.

One of his discoveries concerned amicable numbers, a concept dating back to ancient times. It was the Greeks who defined two whole numbers to be *amicable* if each was the sum of the proper divisors of the other. The numbers 220 and 284 are an example. That is, the divisors of 220 are 1, 2, 4, 5, 10, 11, 20, 22, 44, 55, 110, and, of course, 220. Discarding the last, we find that the sum of the *proper* divisors of 220 is

$$1 + 2 + 4 + 5 + 10 + 11 + 20 + 22 + 44 + 55 + 110 = 284$$

On the other hand, upon adding the proper divisors of 284, we get

$$1 + 2 + 4 + 71 + 142 = 220$$

Thus 220 and 284 are an amicable pair.

For many centuries this was the only known example of amicable numbers. The next breakthrough came from the thirteenth-century Arabic mathematician ibn al-Banna, who discovered the considerably more complicated pair 17,296 and 18,416.[5] In 1636 the French mathematician Pierre de Fermat (the subject of our next chapter) rediscovered al-Banna's numbers and seemed quite pleased with this achievement. But in 1638, feeling somewhat less than amicable, his fierce rival René Descartes (1596–1650) boasted about having found the even more stupendous pair 9,363,584 and 9,437,056. In the slang of today, the message from Descartes to Fermat was, "In your face!"

No further progress was made until Leonhard Euler came along in the eighteenth century. We stress that, at this point, only three sets of amicable pairs were known: those of the Greeks, al-Banna, and Descartes. Euler took a deep breath, went to work, and produced nearly 60 additional pairs. In your face, Descartes!

Of course, what Euler did was to perceive a hitherto unrecognized pattern that allowed him to generate amicable numbers by the bushel. Leonhard Euler could examine an age-old problem and see something that had escaped the finest minds of previous generations.

The same was true in elementary geometry, a field so thoroughly explored that one might expect it to yield no further surprises. But Euler spotted something new. In fact, four large volumes and another 1600 pages of his *Opera Omnia* are devoted to geometry.

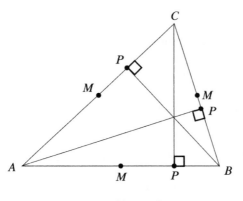

Figure E.1

To see Euler the geometer in action, begin with arbitrary triangle *ABC* as shown in Figure E.1. Bisect each side and draw the triangle's three altitudes; that is, the perpendiculars from each vertex to the opposite side. In the figure we have put an *M* at the midpoint of each side and a *P* at the foot of each perpendicular. What, if anything, is noteworthy about these six seemingly unrelated points?

Euler proved the remarkable fact that all six lie on a single circle![6] The circle's center is found as shown in Figure E.2. Let *D* be the common point where the three altitudes of the triangle intersect (technically called the triangle's **orthocenter**) and let *E* be the common point where the perpendicular bisectors of the three sides intersect (the **circumcenter**). Draw the segment connecting *D* and *E* and bisect this segment at *O*. Then *O* is the center of a circle passing through all six points cited above.[7] This extremely peculiar theorem, overlooked by Euclid, Archimedes, Ptolemy, and everyone else for thousands of years, shows that Euler could geometrize with the best of them.

What about calculus? Did Euler have anything to contribute to this subject? The answer is an emphatic yes. For starters, he wrote texts on both differential and integral calculus that consume almost 2,200 pages of volumes 10–13 of his *Opera Omnia*. These expositions taught calculus to generations of mathematicians and have influenced the subject down to modern times. Today's students, who complain about the size and expense of their 6-pound calculus book should be thankful not to be studying the subject from Euler, whose four massive volumes fill a small suitcase.

One of Euler's early triumphs in this arena involved the summation of a specific infinite series. An unsolved problem dating back to the previous century was to evaluate the series

$$1 + \frac{1}{4} + \frac{1}{9} + \frac{1}{16} + \frac{1}{25} + \frac{1}{36} + \frac{1}{49} + \ldots + \frac{1}{k^2} + \ldots$$

in which are summed the reciprocals of the squares of all whole numbers.

For some time mathematicians had known that this series adds to (mathematicians say "converges to") a finite sum. But *what* sum? Even Leibniz, the creator of

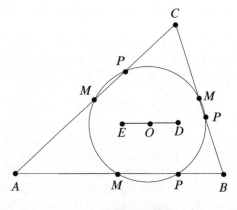

Figure E.2

calculus, was clueless. So, too, were the Bernoulli brothers. Both Jakob and Johann Bernoulli worked on the problem, not only for its intrinsic interest but also presumably for the bragging rights that would accrue. Imagine the humiliation Jakob could have inflicted upon Johann—or vice versa—by solving so difficult and famous a problem.

But no progress was made until Leonhard Euler gave the series a look in 1734. At first, he, too, was baffled. He resorted to a tedious calculation to show that its sum was approximately 1.6449, but this was certainly not a recognizable number. Euler may have been tempted to join the illustrious company of Leibniz and the Bernoullis in admitting defeat, but then, in his words, "quite unexpectedly I have found an elegant formula . . . depending on the quadrature of the circle."[8]

What he meant was that his solution, which veered off into trigonometry and calculus, required the circular constant π. With a boldness to match his brilliance, Euler proved that

$$1 + \frac{1}{4} + \frac{1}{9} + \frac{1}{16} + \frac{1}{25} + \frac{1}{36} + \frac{1}{49} + \ldots = \pi^2/6$$

Although we must refer the reader to the chapter notes for details of the proof, suffice it to say that the sum of $\pi^2/6$ came as a complete surprise to everyone.[9] By resolving a problem that had so thoroughly stumped his predecessors, Euler put the European mathematical community on notice that a new star had risen.

There is much more we could say about Euler's mathematical achievements. But we conclude this chapter by presenting, in some detail, a single theorem from 1740. This example, a mere crumb from Euler's mathematical table, serves as a representative to illustrate the power so typical of his work.

The problem in question was suggested to Euler in a letter from the French mathematician Philippe Naudé. In the autumn of 1740 Naudé inquired about the number of ways a positive integer could be written as the sum of distinct positive integers. The matter caught Euler's interest. He sent an answer within days, accompanied by an apology for a delay caused by "bad eyesight for which I have been suffering for some weeks."[10]

Before examining Euler's proof, we should take a quick look at the decomposition of whole numbers into sums of integers. Consider $n = 6$. The following are the four ways of writing 6 as the sum of *distinct* whole numbers:

$$6, 5 + 1, 4 + 2, 3 + 2 + 1$$

(Here we regard the "sum" consisting of 6 alone as an allowable representation.) Of course, the decomposition $6 = 3 + 3$ is impermissible because the summands are not distinct. Note also that we do not restrict the number of summands: We can have one, two, or more numbers, provided they are all different and add to 6. Introducing the notation $D(n)$ to stand for the number of ways of writing n as the sum of distinct whole numbers, we have shown that $D(6) = 4$.

Now consider the number of ways of writing 6 as the sum of *odd* whole numbers, where we no longer insist that these summands be different. A moment's experimentation leads to the following possibilities:

$$5 + 1, 3 + 3, 3 + 1 + 1 + 1, 1 + 1 + 1 + 1 + 1 + 1$$

Notice that here repetitions are permitted, although we are now restricted to odd summands only. Letting $O(n)$ be the number of ways of writing n as the sum of (not necessarily distinct) odd integers, we see that $O(6) = 4$.

Is it a mere coincidence that there are as many ways to write 6 as the sum of distinct integers as there are to write it as the sum of odd integers? The obvious next step is to repeat the process with a different number. Such a practice is common among mathematicians who, like chemists, gain valuable insight by experimenting with specific cases before trying to formulate and prove general laws. Of course, unlike chemists, mathematicians need not worry about being blown up by their experiments.

Take a look at $n = 13$. The 18 ways to write 13 as the sum of distinct whole numbers are:

13	8 + 4 + 1
12 + 1	8 + 3 + 2
11 + 2	7 + 5 + 1
10 + 3	7 + 4 + 2
9 + 4	7 + 3 + 2 + 1
8 + 5	6 + 5 + 2
7 + 6	6 + 4 + 3
10 + 2 + 1	6 + 4 + 2 + 1
9 + 3 + 1	5 + 4 + 3 + 1

In our notation, $D(13) = 18$. Likewise, the following are the ways of writing 13 as the sum of odd integers:

13	$5 + 5 + 1 + 1 + 1$
$11 + 1 + 1$	$5 + 3 + 3 + 1 + 1$
$9 + 3 + 1$	$5 + 3 + 1 + 1 + 1 + 1 + 1$
$9 + 1 + 1 + 1 + 1$	$5 + 1 + 1 + 1 + 1 + 1 + 1 + 1 + 1$
$7 + 5 + 1$	$3 + 3 + 3 + 3 + 1$
$7 + 3 + 3$	$3 + 3 + 3 + 1 + 1 + 1 + 1$
$7 + 3 + 1 + 1 + 1$	$3 + 3 + 1 + 1 + 1 + 1 + 1 + 1 + 1$
$7 + 1 + 1 + 1 + 1 + 1 + 1$	$3 + 1 + 1 + 1 + 1 + 1 + 1 + 1 + 1 + 1 + 1$
$5 + 5 + 3$	$1 + 1 + 1 + 1 + 1 + 1 + 1 + 1 + 1 + 1 + 1 + 1 + 1$

A quick count reveals that $O(13) = 18$ also. This suggests that we may be onto something.

It is easy to imagine Euler, with his talent for calculation, trying more examples. Each experiment yielded the same surprising result: The number could be decomposed into distinct integers in precisely as many ways as it could be decomposed into odd integers.

Euler noticed this phenomenon, but he did much more: In a single master stroke he *proved* the equality of $D(n)$ and $O(n)$ for all whole numbers. His method of attack, involving a small bit of algebra and a large dose of ingenuity, can be resolved into three simple steps. Looking over Euler's shoulder, we begin with

STEP 1: Introduce the infinite product.

$$P(x) = (1 + x)\,(1 + x^2)\,(1 + x^3)\,(1 + x^4)\,(1 + x^5)\,(1 + x^6)\ldots$$

where the terms continue in the obvious pattern. Never timid, Euler multiplied out and collected like powers of x. The constant term of the resulting expression is clearly 1, formed by multiplying an infinitude of 1s. There is only one way to get an x from such a multiplication, namely when the x in the first factor multiplies only 1s thereafter. Likewise x^2 occurs only once. But there are two ways to get an x^3: One when the x^3 from the third factor multiplies all 1s and the other when the x from the first term multiplies the x^2 from the second. We write this as $(x^3 + x^{2+1})$ to indicate not only the two x^3 terms but also the manner in which they arise. With this convention, the two x^4 terms are expressed as $(x^4 + x^{3+1})$, the three x^5 terms as $(x^5 + x^{4+1} + x^{3+2})$, the four x^6 terms as $(x^6 + x^{5+1} + x^{4+2} + x^{3+2+1})$, and so on.

The process can be continued until we grow weary, but the pattern is evident: There are as many x^n terms in the expansion of $P(x)$ as there are ways to write n as the sum of distinct positive integers. In this regard, note that the exponents on the four x^6 terms are precisely the four decompositions of 6 into distinct integers that we found above. Distinctness is guaranteed because each factor in the expression for $P(x)$ contains a *different* power of x.

Therefore, the coefficient of x^n in the expansion of $P(x)$ is just $D(n)$, the number of decompositions of n into distinct factors. In other words,

$$P(x) = 1 + D(1)x + D(2)x^2 + D(3)x^3 + D(4)x^4 + D(5)x^5 + \ldots + D(n)x^n + \ldots$$

STEP 2: Setting $P(x)$ aside for the time being, introduce the expression

$$Q(x) = \left(\frac{1}{1-x}\right)\left(\frac{1}{1-x^3}\right)\left(\frac{1}{1-x^5}\right)\left(\frac{1}{1-x^7}\right)\ldots$$

where the denominators feature all odd powers of x in increasing order. Euler first had to transform each of these fractions into equivalent nonfractional expressions.

But how? Disregarding the subtleties of infinite series, we observe that

$$1 = 1 - a + a - a^2 + a^2 - a^3 + a^3 - a^4 + a^4 - \ldots$$

because on the right side all terms but the first will cancel. Pairing off the terms on the right and factoring allows us to convert this to

$$1 = (1 - a) + (a - a^2) + (a^2 - a^3) + (a^3 - a^4) + (a^4 - a^5) \ldots$$

$$= (1 - a) + a(1 - a) + a^2(1 - a) + a^3(1 - a) + a^4(1 - a) + \ldots$$

Then divide both sides by $1 - a$ to get

$$\frac{1}{1-a} = \frac{1-a}{1-a} + \frac{a(1-a)}{1-a} + \frac{a^2(1-a)}{1-a} + \frac{a^3(1-a)}{1-a} + \frac{a^4(1-a)}{1-a} + \ldots$$

$$= 1 + a + a^2 + a^3 + a^4 + \ldots$$

because we can cancel $(1 - a)$ from each fraction in the string. In this way we conclude:

$$\frac{1}{1-a} = 1 + a + a^2 + a^3 + a^4 + \ldots \tag{*}$$

Now replace a by x in the formula to get

$$\frac{1}{1-x} = 1 + x + x^2 + x^3 + x^4 + \ldots$$

which we write equivalently as

$$\frac{1}{1-x} = 1 + x^1 + x^{1+1} + x^{1+1+1} + x^{1+1+1+1} + \ldots$$

Next substitute x^3 for a in formula (*) to get

$$\frac{1}{1-x^3} = 1 + (x^3) + (x^3)^2 + (x^3)^3 + (x^3)^4 + \ldots$$

$$= 1 + x^3 + x^{3+3} + x^{3+3+3} + x^{3+3+3+3} + \ldots$$

Similarly, substituting x^5 for a in (*) yields

$$\frac{1}{1-x^5} = 1 + (x^5) + (x^5)^2 + (x^5)^3 + (x^5)^4 + \ldots$$

$$= 1 + x^5 + x^{5+5} + x^{5+5+5} + x^{5+5+5+5} + \ldots$$

and so on.

With these expressions, Euler transformed $Q(x)$ as follows:

$$Q(x) = \left(\frac{1}{1-x}\right)\left(\frac{1}{1-x^3}\right)\left(\frac{1}{1-x^5}\right)\left(\frac{1}{1-x^7}\right)\cdots$$

$$= (1 + x^1 + x^{1+1} + x^{1+1+1} + x^{1+1+1+1} + \ldots)$$

$$(1 + x^3 + x^{3+3} + x^{3+3+3} + \ldots)(1 + x^5 + x^{5+5} + \ldots)\ldots$$

and from here he multiplied out. Again, there emerges a constant term of 1, followed by a single $x^1 = x$ and a single $x^{1+1} = x^2$. There will be two occurrences of the cubic term: x^3 and x^{1+1+1}. We get an x^4 when the x^3 in the second parenthesis multiplies the x^1 in the first one (with each remaining parenthesis contributing a 1), and we get another x^4 when $x^{1+1+1+1}$ in the first multiplies only 1s elsewhere. Similarly, x^5 arises three times (x^5, x^{3+1+1}, and $x^{1+1+1+1+1}$). There are four occurrences of x^6, namely x^{5+1}, x^{3+3}, $x^{3+1+1+1}$, and $x^{1+1+1+1+1+1}$.

And on it goes. It should be clear that the number of x^n terms appearing in the expansion of $Q(x)$ will be precisely the number of ways of writing n as the sum of *odd* integers, as only odd powers occur as exponents in $Q(x)$. As additional verification of this principle, note that the four exponent arrangements for x^6 are the four ways of writing 6 as the sum of odd integers that we saw earlier. Therefore, when $Q(x)$ is expressed as an infinite sum, the coefficient of x^n is $O(n)$ as defined above. That is,

$$Q(x) = 1 + O(1)x + O(2)x^2 + O(3)x^3 + O(4)x^4 + O(5)x^5 + \ldots + O(n)x^n + \ldots$$

STEP 3: Having rewritten $P(x)$ and $Q(x)$ in steps 1 and 2, we finally will show the quite unexpected fact that these two are actually the same thing. To do so, begin with the original expression for $Q(x)$ and multiply numerator and denominator by the product of $(1 - x^2)$, $(1 - x^4)$, and all other even-powered terms. That is,

$$Q(x) = \left(\frac{1}{1-x}\right)\left(\frac{1}{1-x^3}\right)\left(\frac{1}{1-x^5}\right)\left(\frac{1}{1-x^7}\right)\cdots$$

$$= \frac{(1-x^2)(1-x^4)(1-x^6)(1-x^8)\ldots}{(1-x)(1-x^2)(1-x^3)(1-x^4)(1-x^5)(1-x^6)(1-x^7)\ldots}$$

Next recall that the expression $(1 - x^2)$ can be factored into the product of $(1 - x)$ and $(1 + x)$; that $(1 - x^4) = (1 - x^2)(1 + x^2)$; that $(1 - x^6) = (1 - x^3)(1 + x^3)$; and so on. Making these replacements in the numerator yields

$$Q(x) = \frac{(1-x)(1+x)(1-x^2)(1+x^2)(1-x^3)(1+x^3)(1-x^4)(1+x^4)\ldots}{(1-x)(1-x^2)(1-x^3)(1-x^4)(1-x^5)(1-x^6)(1-x^7)\ldots}$$

from which we see at once that each term in the denominator cancels its mate in the numerator. When this wholesale cancellation is complete, there remains

$$Q(x) = (1+x)(1+x^2)(1+x^3)(1+x^4)\ldots$$

which, as if by magic, is precisely the formula for $P(x)$ with which we began. In short, $Q(x)$ and $P(x)$ are indeed equal.

But now we quickly follow Euler to his conclusion. Because we previously established that

$$P(x) = 1 + D(1)x + D(2)x^2 + D(3)x^3 + D(4)x^4 + D(5)x^5 + \ldots + D(n)x^n + \ldots$$

and that

$$Q(x) = 1 + O(1)x + O(2)x^2 + O(3)x^3 + O(4)x^4 + O(5)x^5 + \ldots + O(n)x^n + \ldots$$

and because in step 3 we found that $P(x)$ and $Q(x)$ are the same, it follows that the individual coefficients of each x^n term must likewise be same. Hence, $D(1) = O(1)$, $D(2) = O(2)$, and generally for any whole number n, $D(n) = O(n)$. In other words, this means there are exactly as many ways of writing n as the sum of *distinct* whole numbers as there are ways of writing it as the sum of (not necessarily distinct) *odd* numbers. This is the conclusion Euler sought, and his proof is complete. ■

This argument, proving a subtle and nonobvious fact about the decompositions of whole numbers, is a masterpiece. It displays the following characteristics that typify Euler's mathematics:

1. He was remarkably adept at manipulating symbolic expressions. This ability is certainly evident in this proof and has earned him a reputation as the greatest symbol manipulator of all times.

2. Euler's talent in juggling algebraic expressions went hand in hand with his faith that such juggling would lead to valid conclusions. Subsequent mathematicians showed that uncritically manipulating symbols, especially those involving infinite processes, can lead to trouble. But Euler seemed to believe devoutly that if we follow the symbols, they will lead to the truth.

3. One of Euler's most fruitful mathematical strategies was to write the same expression in two different ways, equate these alternate representations, and from them draw powerful conclusions. This was certainly the case in our example, where $P(x)$ and $Q(x)$ offered alternative ways to express the same thing. The ability to look at one object from two radically different perspectives characterizes many of Euler's most profound and beautiful arguments.

4. Finally, when the algebraic manipulation and technical prowess are stripped away, there remains a degree of ingenuity that is simply astonishing. In the proof above, what insight suggested that to gather information about the decomposition of whole numbers one should expand algebraic expressions? What insight led him to the expressions $P(x)$ and $Q(x)$? And what insight showed him that these would be equal? After one understands his proof, there is a temptation to pronounce it as self-evident. Such is the advantage of hindsight. But it took a remarkable intellect to blaze a way through this uncharted territory.

We should add a final word. Leonhard Euler was a mathematician of the very first rank, yet he is almost universally unknown among the general public, most of whom presumably cannot even correctly pronounce his name. The same people who have never heard of Euler would have no trouble identifying Pierre-Auguste Renoir as an artist or Johannes Brahms as a musician or Sir Walter Scott as an author. Euler's contrasting anonymity is both an injustice and a shame.

But what makes it all the worse is that Euler's counterpart among painters is not Renoir but Rembrandt; his counterpart among musicians is not Brahms but Bach; and his counterpart among writers is certainly not Walter Scott but William Shakespeare. That a mathematician with such peers—the Shakespeare of mathematics—commands so little public recognition is a sad, sad commentary.

So, readers are urged to toss this book aside and begin forming fan clubs, making banners, and otherwise spreading the word about one of the most insightful, most influential, and most ingenious mathematicians of them all: Leonhard Euler of Switzerland.

ermat

$$x^n + y^n \neq z^n$$

Any biographical sketch of Pierre de Fermat (1601–1665) is bound to be short. His life spanned the first two-thirds of the seventeenth century but was, truth to tell, rather dull. He never held an appointment at a university nor occupied a chair at a royal academy. By training a lawyer, by profession a magistrate, Fermat published almost nothing during his lifetime, instead conveying his ideas through correspondence and unpublished manuscripts. Because he was not a professional mathematician, Fermat has been dubbed the "prince of amateurs." But if by "amateur" we mean "marginally talented novice," the moniker is totally inaccurate.

"Mathematical amateur"—the term has a strange ring to it. Were we to classify people as either professional or amateur mathematicians, then virtually everyone in history would fall into the latter category. Accordingly, addition errors in your checkbook would properly be classified as "the mathematics of amateurs." So, too, would Yogi Berra's observation that "Success is 90% hard work and the other 20% luck."*

Such statements are light years removed from the mathematical output of the "amateur" Fermat. And if he is less well known to the general public than his two great French contemporaries, René Descartes and Blaise Pascal, he occupies a more esteemed place in the hearts of mathematicians. The primary aim of this chapter is to explain why.

Pierre Fermat was born in Beaumont de Lomagne in southern France at the beginning of the seventeenth century. His father was a well-to-do merchant and

---

*Who knows where Yogi gets these things?

town consul, and in comfortable circumstances young Pierre spent his childhood. He received a fine education heavily slanted toward the study of classical languages and literatures and subsequently attended university to concentrate on legal matters. This training led to his service as magistrate in the *parlement* of the city of Toulouse, a post that, besides providing financial security, gave Fermat the right to insert the "de" before his last name to signify a kind of minor French nobility.

As a figure of prominence, Fermat married and, with his wife, raised five children. He held positions of influence in the Catholic Church, of which he was a devout member. From what we know, he lived his entire life within a hundred miles of his birthplace.[1] This Frenchman never saw Paris.

In short, Pierre de Fermat led a restrained and quiet existence—so quiet, in fact, that he may not have had enough to do. It has been suggested that the undemanding nature of his job provided time to write Latin poetry or scholarly critiques of Greek texts. With excess time and a penetrating intelligence, Fermat brings to mind the young Albert Einstein some two and a half centuries later, whose unchallenging position as a Swiss patent clerk gave him the opportunity to develop the theory of relativity.

Fermat's real passion—more intense than classical poetry, church business, or even the law—was mathematics, where his contributions were both profound and far ranging. He played significant roles in the development of many of the topics thus far addressed in this book: number theory, probability, differential calculus. As noted, he eschewed publication of his mathematical discoveries, perhaps for reasons suggested by his remark, "I have so little aptitude in writing out my demonstrations that I have been content to have discovered the truth, and to know the means of proving it when I shall have the occasion to do so."[2]

Fortunately he communicated his ideas by letter to other scholars around Europe. In this, the jurist from Toulouse was a tireless correspondent, and his letters provide our best information about his mathematics. The recipients of these letters— Descartes, Pascal, Christiaan Huygens, John Wallis, and Marin Mersenne—read like a "who's who" of science spanning the first half of the seventeenth century. From them Fermat could learn what was going on in Paris or Amsterdam or Oxford; to them he could convey his own remarkable discoveries.

Among the most impressive of these were his formulation of what we now call analytic geometry, described in a 1636 treatise titled *Ad locus planos et solidos isagoge*, and his contributions to the foundations of probability theory contained in a series of letters from 1654. With respect to the latter, Fermat's name is linked with that of his collaborator, Blaise Pascal (1623–1662). In their extensive correspondence, they outlined ideas, offered criticism, and nudged the hitherto overlooked theory of probability into the mathematical limelight. Much of their joint work would find its way, directly or indirectly, into the *Ars Conjectandi* of Jakob Bernoulli described in Chapter B.

As for analytic geometry, Fermat's name is also linked with another mathematician, although this time not as a collaborator. The other was René Descartes, who independently devised his own system of analytic geometry. Both individuals

Pierre de Fermat
(Courtesy of Lafayette College Library)

grasped the enormously fertile idea of combining two great currents of mathematical thought: the geometric and the algebraic. (See Chapter XY for further discussion of this topic.)

Unfortunately but characteristically, Fermat never published his treatise, whereas Descartes told the world about his findings in the influential *Géométrie* of 1637. As a consequence of being first in print, it was Descartes who received the public acclaim and whose name has ever after been enshrined in the term *Cartesian plane*. Had our French magistrate been a little more forthcoming, mathematicians might instead talk of the Fermatian plane.

Descartes won this battle but certainly not the war. In fact, his interest in mathematics was less consuming than that of Fermat, who made so many other significant contributions that his cocreation of analytic geometry is often overlooked. One such contribution, and an instance in which Fermat clearly bested Descartes, was in finding maximum and minimum values of certain curves.

This should sound familiar. It was one of the key objectives of differential calculus as discussed in Chapter D. There we gave credit to Leibniz and Newton for formulating the procedures necessary to determine extreme values, but we neglected to mention that decades earlier Fermat had devised very similar methods. These appeared in his *Methodus ad disquirendam maximum et minimam*, another brilliant but (typically) unpublished work.

His treatment of maxima, minima, and tangents brought Fermat into conflict with Descartes during the closing years of the 1630s. The latter had developed his own technique to handle problems of tangents and asserted that "this is not only the most useful and most general problem in geometry that I know, but even that I have ever desired to know."[3] However, Descartes's approach proved cumbersome even for elementary examples. What Fermat could do almost effortlessly required Descartes to fill page after page with gruelling algebraic computations.

For a time, a rivalry developed over the matter, with Descartes proclaiming the superiority of his own method. However, it soon became apparent, even to *him*, that Fermat had taken the better route. The acknowledgment of defeat—a rare experience for René Descartes—left a residual bitterness between the two greatest mathematicians of the age.

Because of the success with which Fermat solved simple maxima and minima problems, Pierre-Simon de Laplace (1749–1827) called him "the true inventor of the differential calculus."[4] This assessment of one French mathematician for another overstates the case, and Laplace appears to have been carried away by an uncontrolled outburst of nationalistic zeal. Fermat's insights notwithstanding, we can cite a number of reasons why he should not be accorded so great an honor.

For one thing, Fermat applied his techniques to a limited family of curves: those of the form $f(x) = x^n$, sometimes called the "parabolas of Fermat," and those of the form $g(x) = 1/x^n$, the "hyperbolas of Fermat." The true inventors of calculus could handle more sophisticated functions so as, in Leibniz's words, "to be impeded neither by Fractional nor Irrational Quantities."

More significantly, Fermat failed to grasp the so-called fundamental theorem of calculus, the subject's great unifying idea that we shall examine in Chapter L. This theorem is so central that a failure to perceive it imposes an automatic disqualification from any claims to have discovered calculus. Both Newton and Leibniz, it should be mentioned, recognized the fundamental theorem quite clearly.

For these reasons, modern historians of mathematics tend not to dub Pierre de Fermat as the creator of calculus. But almost all admit that he came mighty close.

So we credit Fermat with major advances in analytic geometry, differential calculus, and probability theory and acknowledge these as remarkable contributions for an "amateur." But all of this is really prologue. Far more than any of these achievements, Fermat's reputation rests upon his wonderful investigations into the theory of numbers.

As noted in Chapter A, this subject had been examined by Euclid and other classical mathematicians, but it is no exaggeration to say that modern number theory begins with Fermat. For this French student of Greek classics, interest in number theory was fueled by ancient texts. This was especially true of Diophantus' *Arithmetica* from A.D. 250, whose 1621 translation had come to Fermat's attention. He meticulously read this work, often jotting marginal comments upon the pages of his well-thumbed copy.

To Fermat the subject was endlessly fascinating. He was enthralled by, and on intimate terms with, whole numbers and had the uncanny ability to recognize their attributes even as one recognizes old friends. Outwardly Pierre de Fermat was a respected jurist in Toulouse, but inwardly he was a number theorist par excellence.

Here we can touch upon only a few of his discoveries. Of course, mathematicians are at a disadvantage because he left behind almost no proofs. A marginal note here, a tantalizing allusion there—these are about all we have. Later scholars, Euler in particular, tried to reconstruct his thought processes and probable lines of reasoning. But in the words of the twentieth-century mathematician André Weil, "when Fermat asserts that he has a proof for some statement, such a claim has to be taken seriously."[5]

One of his most amazing assertions concerns the decomposition of prime numbers into the sum of two perfect squares. The essence of Fermat's discovery requires a few preliminaries.

First, it is clear that if a whole number n is divided by 4, the remainder is either 0, 1, 2, or 3. Remainders, after all, must be less than divisors. Mathematicians say that any whole number fits into one of four categories:

$n = 4k$ (the number is an exact multiple of 4)

$n = 4k + 1$ (the number is one more than a multiple of 4)

$n = 4k + 2$ (the number is two more than a multiple of 4)

$n = 4k + 3$ (the number is three more than a multiple of 4)

Numbers of the form $4k$ and $4k + 2$ are obviously even, and thus, except for the trivial case of 2, are not prime. Any odd number—and more importantly any odd prime—*must* look like either $4k + 1$ or $4k + 3$.

Needless to say, examples of both types are plentiful. In the former category are primes $5 = 4(1) + 1$, $13 = 4(3) + 1$, and $37 = 4(9) + 1$; in the latter are $7 = 4(1) + 3$, $19 = 4(4) + 3$, and $43 = 4(10) + 3$. All odd primes fall into one or the other category.

Beyond this defining characteristic, the two types of odd primes may seem roughly equivalent. In fact, they differ in a significant and surprising respect, and this difference is at the heart of Fermat's assertion.

The year was 1640. Fermat, in a Christmas letter to Father Mersenne, made the following observation: "A prime number, which exceeds a multiple of four by unity, is only once the hypotenuse of a right triangle."[6] This was his curiously geometric way of stating that primes in the first category—those of the form $4k + 1$—could be decomposed into the *sum* of two perfect squares and that this decomposition could be achieved in one and only one way. On the other hand, he observed that primes of the form $4k + 3$ could not be written as the sum of two perfect squares in any way at all. In this light, the two categories of odd primes appear remarkably dissimilar: One is the "sum of two squares" type and the other is not.

Some examples might be helpful. For the prime $13 = 4(3) + 1$, we have the decomposition $13 = 4 + 9 = 2^2 + 3^2$. For $37 = 4(9) + 1$, we can write $37 = 1 + 36 = 1^2 + 6^2$; and for the more challenging example of prime $193 = 4(48) + 1$, we have $193 = 49 + 144 = 7^2 + 12^2$. By contrast, primes such as $19 = 4(4) + 3$ or $199 = 4(49) + 3$ have no decomposition into the sum of two perfect squares.

It is not difficult to prove the latter fact. We need only examine what happens to even and to odd numbers when they are squared.

THEOREM: An odd number of the form $n = 4k + 3$ cannot be written as a sum $a^2 + b^2$ of two perfect squares.

PROOF: We consider the three possible cases.

Case 1: If both a and b are even, so are a^2 and b^2. Hence $a^2 + b^2$, the sum of two even numbers, is itself even and thus could not equal the *odd* number $n = 4k + 3$.

Case 2: If both a and b are odd, their squares a^2 and b^2 are odd as well. Thus $a^2 + b^2$, being the sum of two odds, is even and could not be $n = 4k + 3$ either.

Case 3: The only remaining possibility is that we add the square of an even number to the square of an odd one. Suppose a is even; this means it is a multiple of 2, so we can write $a = 2m$ for some whole number m. Because b is odd, it is one more than a multiple of 2, and we write $b = 2r + 1$ for some whole number r. So when we sum the squares of an even and an odd we get:

$$a^2 + b^2 = (2m)^2 + (2r + 1)^2$$
$$= 4m^2 + (4r^2 + 4r + 1) \quad \text{by the rules of algebra}$$
$$= 4(m^2 + r^2 + r) + 1 \quad \text{by factoring out the common 4}$$

Thus $a^2 + b^2$ is one more than $4(m^2 + r^2 + r)$—that is, one more than a multiple of 4. Although this indeed makes $a^2 + b^2$ an odd number, it could not possibly be $n = 4k + 3$, which is *three* more than a multiple of 4.

To summarize, if a and b are both even or both odd, the expression $a^2 + b^2$ is even; if a is even and b is odd (or vice versa), the expression $a^2 + b^2$ is one more than a multiple of 4. In no case could $a^2 + b^2$ be three more than a multiple of 4. Thus odd numbers, and certainly odd primes, of the form $4k + 3$ cannot be written as the sum of two perfect squares. The proof is complete. ■

So much for half of Fermat's result. The other half—that primes of the form $4k + 1$ *can* be written as the sum of two perfect squares in one and only one fashion—is far too difficult to be proved here. As usual, Fermat left only a vague and tantalizing indication of his argument. It was Euler who first published the proof a little more than a century later.[7]

The dichotomy among primes that lies at the heart of the theorem is both intriguing and nonintuitive. To see a bit of its power, consider the problem of determining whether the number $n = 53,461$ is a prime. (Recall from Chapter A that Gauss characterized the question of determining primality as "one of the most important and useful in arithmetic.") A little checking reveals that easy candidates for prime divisors—numbers such as 2, 3, 5, 7, 11, 13—do not work, and we may quickly tire of the search for factors.

But observe three telling facts:

   a. $n = 53,461 = 4(13,365) + 1$, so n has the form $4k + 1$.

   b. $n = 53,461 = 100 + 53,361 = 10^2 + 231^2$, so n can be written as the sum of two squares in one way.

   c. $n = 53,461 = 11,025 + 42,436 = 105^2 + 206^2$, so n can be written as the sum of two squares in a second way.

From these three clues we deduce that n is composite. Otherwise, it would be a prime of the form $4k + 1$ having two *different* decompositions into the sum of squares as exhibited in facts (b) and (c). Such a situation is impossible by the theorem of Fermat just stated. Hence n cannot be prime.

Two features of this argument might be cause for reader annoyance. First, it is fair to ask how one comes up with the decompositions into squares of facts (b) and (c); determining these decompositions appears every bit as hard as factoring the original number. In response to accusations that this example was contrived, we plead guilty: It *was* contrived.

More important is the other source of possible consternation, namely our willingness to conclude that $n = 53,461$ is not prime *without exhibiting a single divisor*. It seems that, to show a number is composite, it is incumbent upon us to produce explicit factors. The approach above, in which we showed only that 53,461 violates a condition that primes possess, has a certain indirectness about it. But our conclusion remains perfectly sound. It is not a flaw in our reasoning but a reminder that there are many weapons, some of them quite subtle, in the number theorist's arsenal.

To set uneasy minds at rest, we now determine the prime factors of 53,461. In so doing, we illustrate yet another of Fermat's discoveries, his ingenious factorization scheme.

Suppose we wish to factor a whole number n. To split n in half "additively," we would of course use $n/2$, because $n/2 + n/2 = n$. But to halve n "multiplicatively," we look at $\sqrt{n}$, because $\sqrt{n} \times \sqrt{n} = n$.

This is where Fermat began his search for factors. Of course $\sqrt{n}$ is rarely a whole number, so we let m be the smallest integer greater than or equal to $\sqrt{n}$. For instance, when trying to factor $n = 187$, we note that $\sqrt{n} = \sqrt{187} \approx 13.67$ and so use $m = 14$.

Now consider the sequence of numbers $m^2 - n$, $(m + 1)^2 - n$, $(m + 2)^2 - n$, etc., and suppose one of these turns out to be a perfect square. That is, suppose we eventually find a number b such that $b^2 - n = a^2$.

A bit of rearrangement converts this to $n = b^2 - a^2$, an important pattern known as "the difference of squares." We factor this difference to conclude

$$n = b^2 - a^2 = (b - a)(b + a)$$

thereby providing an algebraic recipe for decomposing n into the two factors $b - a$ and $b + a$.

As a simple example of Fermat's scheme, we shall factor 187. Begin with $m = 14$ and look at $14^2 - 187 = 196 - 187 = 9 = 3^2$. On our very first try, we got a perfect square. Hence, with $b = 14$ and $a = 3$, we have

$$187 = 14^2 - 9 = 14^2 - 3^2 = (14 - 3)(14 + 3) = 11 \times 17$$

and 187 has been factored.

Sufficiently warmed up, we now have a go at $n = 53{,}461$. Note that $\sqrt{n} = \sqrt{53461} \approx 231.216$, so we start with $m = 232$ and embark upon the search for a perfect square:

$232^2 - 53{,}461 = 363$, not a perfect square

$233^2 - 53{,}461 = 828$, not a perfect square

$234^2 - 53{,}461 = 1295$, not a perfect square (although $1296 = 36^2$)

$235^2 - 53{,}461 = 1764 = 42^2$. Success!

Hence $53{,}461 = 235^2 - 42^2 = (235 - 42)(235 + 42) = 193 \times 277$, and our number, with help from Fermat, has been factored into primes. It was, all in all, a fairly painless procedure, especially when compared to a trial-and-error search. It shows that a little ingenuity goes a long way.

We conclude this chapter with what is, for better or worse, the most famous number theoretic statement from the pen of Fermat. Its renown comes in spite of—or, more probably because of—its extreme difficulty. We are describing what has come to be known as "Fermat's last theorem."

Characteristically, the story begins with Fermat poring over a Greek text, Diophantus' *Arithmetica*, where the topic at hand was again the sum of two perfect squares. Such a sum can, in certain cases, be a square itself. Examples that come to mind are $3^2 + 4^2 = 5^2$ or $420^2 + 851^2 = 949^2$ (admittedly, the former comes to mind a bit more quickly than the latter). But, Fermat mused, can the sum of two perfect *cubes* be yet another perfect cube?

At this point, in the margin of his copy of *Arithmetica*, he wrote, "To divide a cube into two other cubes, or a fourth power, or in general any power whatever into two powers of the same denomination above the second is impossible."[8] In symbols, Fermat was asserting that there are no positive integers x, y, and z such that $x^3 + y^3 = z^3$, or $x^4 + y^4 = z^4$, or $x^5 + y^5 = z^5$, and so on. His general claim was that, if $n \geq 3$, the equation $x^n + y^n = z^n$ has no whole number solutions x, y, and z.

As if to tantalize generations of scholars, Fermat added what is perhaps the most famous statement in all of mathematics: "I have assuredly found an admirable proof of this, but the margin is too narrow to contain it."[9]

Here, then, is his doubly misnamed "last theorem." It is "last" not because it was the final statement of Fermat's career but because it remained in doubt after all of his other assertions were resolved. Moreover, it is a misnomer to call it his "theorem," for he furnished no proof.

We observe that all it would take to disprove Fermat's conjecture is a single exponent $n \geq 3$ and three specific numbers x, y, z for which $x^n + y^n = z^n$. Of course, no one has ever found such numbers.

On the other hand, to prove the conjecture it is necessary to craft an argument valid for *all* exponents $n \geq 3$, and this, to say the least, has presented problems. Certain cases were readily handled. Fermat himself is credited with a proof that no positive integers exist for which $x^4 + y^4 = z^4$. During the eighteenth century Euler gave a substantially correct proof that $x^3 + y^3 = z^3$ was likewise impossible, although he made a prescient observation that the separate arguments for cubes and fourth powers differed so radically as to give no inkling of how to tackle the general theorem.

As the decades passed, other mathematicians were caught up in the quest. Sophie Germain (1776–1831) made important contributions—much too intricate to be described here—that channeled subsequent efforts in well-specified directions. In 1825, the youthful P. G. Lejeune Dirichlet (1805–1859) and the grandfatherly A. M. Legendre (1752–1833) showed that two fifth powers cannot sum to a fifth power. In 1832, Dirichlet eliminated the possibility that $x^{14} + y^{14} = z^{14}$, and a few years later Gabriel Lamé (1795–1870) disposed of $x^7 + y^7 = z^7$. Then in 1847, Ernst Kummer (1810–1893) developed a powerful strategy to show that Fermat's conjecture was true for a large class of exponents. Of course, this in no way eliminated the possibility that it might also be *false* for a large class of exponents.[10] Things were moving slowly.

Interest in the problem continued into the twentieth century, fueled in part by the 1909 offer of 100,000 German marks for a correct solution. The prospect of financial gain brought out the worst in a number of avaricious pseudo-mathematicians, and a flood of erroneous arguments washed over the scholarly world. In an afternote to E. T. Bell's *The Last Problem* (recently revised and updated by Underwood Dudley), there is an amusing anecdote about a mathematician whose response to the onslaught of fallacious proofs was a form letter that began:

Dear Sir or Madam:

Your proof of Fermat's Last Theorem has been received. The first mistake is on page ____, line ____ .[11]

With the astronomical inflation rates of post–World War I Germany, the prize was devalued to the point of absurdity, and there certainly were easier ways to come up with 100,000 marks.

Fortunately mathematicians are not always driven by financial incentive. One with a loftier motive was Gerd Faltings (1954–). In 1983 Faltings proved that for any $n \geq 3$, Fermat's equation $x^n + y^n = z^n$ can have at most finitely many different solutions (apart from those in which one set of solutions is just a multiple of another). At first glance, this may not seem terribly helpful. Faltings did not rule out the possibility that, for some exponents, the equation has 100,000 solutions, which is a far cry from Fermat's insistence that it has *none*. Still, Faltings put a cap on the abundance of solutions for the general case. For his proof he received the 1986 Fields Medal, the mathematical equivalent of the Nobel Prize, at the International Congress of Mathematicians held in Berkeley, California.

As this book goes to press, mathematicians are abuzz over a promising new proof of Fermat's last theorem from Dr. Andrew Wiles of Great Britain. Excitement is running so high that the story landed on the front page of *The New York Times* and was deemed sufficiently newsworthy to warrant a full page article in *Time* (where stories about mathematics are as rare as ads for *Newsweek*).[12] If Wiles's proof withstands the forthcoming scrutiny of the mathematical community, it will be an extraordinary triumph and his name will be writ large in math history books yet to come. If his proof is found to contain an error, it will join thousands of other might-have-beens. Stay tuned.

This may be a good place to leave our unassuming jurist, Pierre de Fermat. Among mathematicians he remains a revered figure, one whose study of the classical masters led him to develop so many pivotal ideas of modern mathematics. In a 1659 letter to a friend, the aging Fermat ventured this hope: "Perhaps posterity will be grateful to me for having shown that the Ancients did not know everything."[13]

We can affirm without the least hesitation that posterity *has* been grateful.

Greek Geometry

Geometry, once the cornerstone of mathematics, was introduced in Chapter C. In this and the following two chapters, we take a deeper look at this ancient and beautiful subject. And what better place to start than with geometry's finest practitioners, the mathematicians of classical Greece?

Greek geometry ranks as a major achievement of the human intellect for reasons both mathematical and historical, both practical and aesthetic. Its golden age stretched from Thales of Miletus of around 600 B.C. down to the second century B.C. work of Eratosthenes, Apollonius, and the unparalleled Archimedes of Syracuse. From there a somewhat less distinguished "silver age" continued to the time of Pappus around A.D. 300. These individuals, and many others, developed geometry from a practical method of measuring land (*geo* = earth, *metria* = measure) to a vast body of abstract theorems and constructions laced together by the uncompromising rules of logic. Greek geometry stands among the great intellectual/artistic movements of Western civilization and in this sense has much in common with Elizabethan drama or French impressionism. Like the impressionists, the Greek geometers shared a general philosophy and style, and though there were as many variants among the Greeks as among the French artists, the deeper, unifying characteristics of an impressionist painting or a Greek theorem are instantly recognizable.

What are these characteristics? Historian Ivor Thomas, in his comprehensive *Greek Mathematical Works*, singles out (1) the impressive logical rigor the Greeks employed for proving theorems, (2) the purely geometric—as opposed to numerical—nature of their mathematics, and (3) their skillful organization in presenting and developing mathematical propositions.[1]

To these characteristics, we add two others. One is their recognition of geometry as an unsurpassed exercise in pure thought, a subject at once ideal, immaterial, and eternal. In the *Republic*, Plato noted that although geometers draw tangible figures as an aid to their investigations,

> they are not thinking about these figures but of those things which the figures represent; thus it is the square in itself and the diameter in itself which are the matter of their arguments, not that which they draw; similarly, when they model or draw objects, which may themselves have images in shadows or in water, they use them in turn as images, endeavoring to see those absolute objects which cannot be seen otherwise than by thought.[2]

Such a view, of course, fits nicely with the Platonic concept of an ideal existence apart from the human experience, and geometrical considerations surely played a role in shaping his philosophy. The Greek thinkers—seeking the perfect, the logical, and the utterly rational—could look to geometry as the embodiment of this ideal.

Less earthshaking, but nonetheless central to much of Greek mathematics, was a reliance on compass and straightedge for geometric constructions. On the one hand, these were two readily available tools for drawing the tangible figures of which Plato talked. But in a more abstract sense, these tools enshrined the straight line (via the straightedge) and the circle (via the compass) as the keys to geometric existence. With the unwavering precision of the ideal straight line and the perfect symmetry of the ideal circle, the Greeks created their geometric figures and, from there, their geometric theorems. And although we today have extended mathematics beyond the limitations of lines and circles, their supremacy for the Greek mathematicians was utterly in character.

There is no question that geometric ideas predated the Greeks. The civilizations of Egypt and Mesopotamia, for instance, used geometry to partition fields and erect pyramids, and we shall return to this topic in Chapter O. But it is to the Greeks that we look for the first geometric theorems, the first propositions proved with logical rigor.

According to tradition, Greece's earliest mathematician, not to mention its earliest astronomer and philosopher, was Thales (pronounced THAY-leez) who grew up along the eastern Aegean shores watching the swimming whales (*not* pronounced WHAY-leez). According to the later commentator Proclus, "Thales was the first to go to Egypt and bring back to Greece this study; he discovered many propositions and disclosed the underlying principles of many others to his successors."[3]

Legend holds that it was Thales who first proved that the base angles of an isosceles triangle are congruent and that any angle inscribed in a semicircle is a right angle (the latter is sometimes given the tongue-twisting name Thales' theorem). Unfortunately, legend is all we have to go on, for his actual proofs disappeared long ago. Still, the ancients held him in very high regard, classifying him as one of the "seven wise men of antiquity." (There is no truth to the rumor that the other six were Grumpy, Happy, Dopey, Sneezy, Doc, and Bashful.)

With Thales, Greek geometry was underway. To trace its subsequent developments, its successes and failures, would occupy countless chapters, if not countless volumes. So here we limit ourselves to two specific geometric issues: how Euclid did geometry with a collapsing compass and why the Epicurean philosophers accused him of being no smarter than an ass. Although these choices may seem a bit odd, they give an accurate sense of the mathematical temperament of the times.

We begin around the year 300 B.C. with Euclid of Alexandria. Although responsible for a number of mathematical treatises, he is best remembered for the *Elements*, a systematic development of much of Greek mathematics up to that point. The work is divided into 13 books and contains 465 propositions on plane and solid geometry and number theory. Aptly called the greatest mathematics textbook of all time, the work has been studied, edited, and revered from its appearance in ancient Greece to the present day.

What makes the *Elements* so important is its logical development from basic principles to sophisticated consequences. Euclid began Book I with a list of 23 definitions so the reader would know precisely what his terms meant. He introduced *point* as "that which has no part" (one of his less illuminating definitions); *equilateral triangle* as a triangle "which has its three sides equal"; and *isosceles triangle* as a triangle "which has two of its sides alone equal."

With terms defined, Euclid presented five postulates to serve as the foundations of his geometry, the starting points from which all would follow. These were given without proof or justification; they were simply to be accepted. Fortunately, such acceptance was not difficult because the postulates appeared to Euclid's contemporaries, and indeed to most of us today, as utterly innocuous. For our purposes in this chapter, we need only the first three:

1. [It is possible] to draw a straight line from any point to any point.

2. [It is possible] to produce a finite straight line continuously in a straight line.

3. [It is possible] to describe a circle with any center and distance.

These seem quite simple, quite self-evident. The first two legitimize the use of the unmarked straightedge in geometric constructions, for they allow us to connect two points with a straight line (postulate 1) or to take an existing line and extend it (postulate 2). This is precisely what straightedges are good for. The third postulate authorizes what we do with a compass: draw a circle with a given point as center and with a predetermined length as radius. Thus, it appears that the first three postulates provide logical support for the operation of the geometric tools.

But the reader may think back to her or his own geometry classes and recall another operation performed with a compass: the transfer of a length from one part of the plane to another. This is easily done. We put the points of the compass at either end of the line segment whose length is to be transferred, lock the compass into place, pick it up, move it rigidly, then set it down at its desired destination. It is a process at once simple and necessary in many geometric constructions.

EUCLID

Taken from a Brass Coin in the Repository of the late Queen Christian of Sweden

Euclid *the Mathematician, was of* Alexandria, *where he taught in the Reign of* Ptolemy Lagus *in the* CXX Olympiad, *and Year of* Rome 454. *He Wrote many things relating to Musick and Geometry: But his* XV Books *of Elements (of which he is generaly thought to be only the Collector) are most applauded: the two last are attributed to* Hypsicles *of* Alexandria, *and not to him*. *Cardan Vossius*.

Yet Euclid included no postulate to justify transferring lengths in this fashion. Where we expect an axiomatic dispensation allowing this procedure, we find nothing. Although his compass could draw a circle, he did not explicitly permit it to be

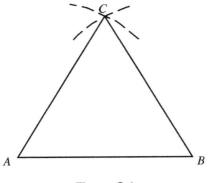

Figure G.1

locked into position and moved about. Euclid's is thus somewhat facetiously called a "collapsible compass," a device that falls shut the instant it is lifted from the page.

This raises a serious question of logic: Did the esteemed Greek geometer forget to include a "transfer of lengths" postulate? Do we have here a Euclidean blunder?

Not at all. As we shall see in a moment, Euclid had his reasons—reasons logically sound and very Greek in nature—for *not* including such a postulate. Rather than a blunder, this omission is evidence of his geometric acumen and organizational ability.

With postulates in place, Euclid introduced a few "common notions": self-evident statements of a more general and less geometric nature. For instance, here we accept without proof that "Things which are equal to the same thing are also equal to one other," that "If equals be added to equals, the wholes are equal," and that "The whole is greater than the part." These are statements with which few will quibble.[4]

Then he was ready to plunge in. With a huge body of geometry to deduce from a tiny collection of definitions, postulates, and common notions, where does one start? This is the sort of initial challenge known to freeze mathematicians (and authors) in their tracks. But if, as the Chinese tell us, a journey of a thousand miles begins with a single step, then Euclid's journey through geometry began with an equilateral triangle. The very first proposition of the *Elements* was the construction, upon a given line segment, of just such a figure.

The argument is easy. Starting with given segment *AB* in Figure G.1, we construct a circle with center at *A* and radius *AB* as permitted by postulate 3. Next, with center *B* and radius *AB*, we invoke the same postulate to construct a second circle. Let *C* be the point where the circular arcs intersect (see the chapter notes regarding the existence of such a point of intersection).[5] We then draw lines *AC* and *BC* by postulate 1, forming △ *ABC*. In this triangle, *AB* and *AC* are the same length because they are radii of the first circle; and *AB* and *BC* are the same length since they are radii of the second. Because things equal to the same thing are equal to each other, all three sides are mutually equal. By Euclid's definition, the triangle is equilateral and the proof is complete.

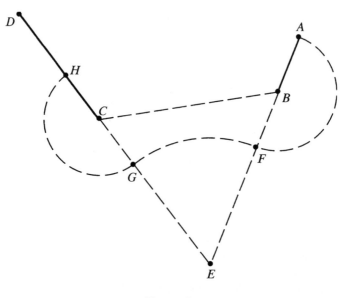

Figure G.2

It is critical to observe that in using the compass for this construction, Euclid never needed to move it rigidly. After each arc was drawn, the compass can fall to pieces without affecting the proof in the least.

But in the next two propositions of Book I, Euclid showed how a length could be transferred *even with a compass that collapses*. This means that length transfer was implied by the postulates already on the table. A new postulate to this end would have been unnecessary baggage, and Euclid was sharp enough to realize this.

His proofs—here combined into a single argument—were quite elegant. Suppose we have segment *AB* as shown in Figure G.2 and wish to transfer its length onto the segment *CD* emanating from the point *C*. First, use the straightedge and apply postulate 1 to draw the segment connecting *B* and *C*. Then upon segment *BC* construct equilateral triangle *BCE*; the legitimacy of this construction, of course, is exactly what was established in the previous proposition.

We then engage in a round of circle drawing. With center *B* and radius *AB*, construct a circle meeting *BE* at *F* (upon picking up the compass, suppose it collapses). With center *E* and radius *EF*, draw a circle meeting *CE* at *G* (again, the compass may collapse as we lift it from the page). With center *C* and radius *CG*, draw a final circle meeting *CD* at *H*. All of these constructions are permitted by Euclid's third postulate; none require a rigid compass.

Now we merely generate a string of equalities (for ease of notation, we shall denote the *length* of segment *XY* by $\overline{XY}$):

$$\overline{AB} = \overline{BF}$$ because they are radii of the same circle

$$= \overline{BE} - \overline{EF}$$

$$= \overline{BE} - \overline{EG}$$ because EF and EG are radii of the same circle

$$= \overline{CE} - \overline{EG}$$ because the three sides of $\triangle BCE$ are equally long

$$= \overline{CG}$$

$$= \overline{CH}$$ because again we have radii of the same circle

The beginning and end of this chain reveal that $\overline{AB} = \overline{CH}$. Therefore the initial length of AB has been transferred onto segment CD as required, yet nowhere did we have to pick up a compass and move it rigidly.

The surprising conclusion of this proof is that constructions seeming to require a noncollapsing compass can actually be accomplished with a collapsible one. As he subsequently developed his geometry, Euclid could therefore legitimately transfer a length from one place to another as if with a rigid compass, his rationale being the just-proved theorem. By getting it out of the way so early, and so simply, he was free to use it in all that followed.

At this point some readers may be stifling a yawn, regarding the whole business as much ado about nothing. After all, everyone knows that stationery stores sell cheap metal compasses that are made to stay open, and surely it would have done no great damage for Euclid to have included an extra postulate to that effect.

Adherents of this position, we believe, have not yet gotten into the spirit of formal Greek geometry. First, the real-world existence of rigid compasses has no bearing on the development of ideal concepts. Second, stationery stores had not yet been invented. Third, and most crucially, Euclid would not have wanted to add an *unnecessary* postulate to his list. Why assume something that could be derived from other assumptions? It would make his postulates less pure, less streamlined, and less perfect and thereby violate an aesthetic, not a mathematical, principle. That aesthetic considerations were critical for the Greek mathematician is obvious. In Euclid's proof above we glimpse what Ivor Thomas meant when he wrote:

> [A] feature which cannot fail to impress a modern mathematician is the perfection of form in the work of the great Greek geometers. This perfection of form, which is another expression of the same genius that gave us the Parthenon and the plays of Sophocles, is found equally in the proof of individual propositions and in the ordering of those separate propositions into books; it reaches its height, perhaps, in the *Elements* of Euclid.[6]

We now move deeper into Book I to see further evidence of Euclid's genius. After disposing of the collapsible compass in propositions 2 and 3, Euclid proved the so-called side-angle-side, or SAS, congruence scheme in proposition 4. That is (see Figure G.3), if we have triangles ABC and DEF in which $\overline{AB} = \overline{DE}, \overline{AC} = \overline{DF}$, and the included angles $\alpha = \delta$, then the triangles are congruent, meaning they have pre-

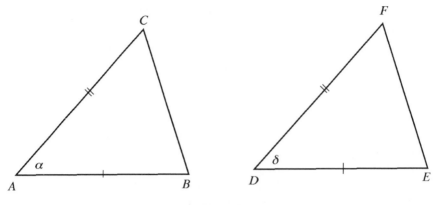

Figure G.3

cisely the same size and shape. In other words, if △ *DEF* were picked up and placed atop △ *ABC*, the two would coincide perfectly, line for line, angle for angle, point for point.

In Euclid's hands, triangle congruence was *the* key to proving geometric propositions. Later he established the additional congruence patterns of side-side-side (or SSS) in proposition 8, and angle-side-angle (ASA) and angle-angle-side (AAS) in proposition 26.

Proposition 5 of Book I proved that the base angles of an isosceles triangle are congruent. As noted, this result has been attributed to Thales, but the proof in the *Elements* is probably Euclid's own.[7] Although we shall not give it here, we note that it was accompanied by the diagram in Figure G.4. This configuration, which suggests a bridge (at least to those with vivid imaginations) may account for calling proposition 5 the *pons asinorum*, or bridge of asses. According to tradition, dullards—that is, asses—find the proof beyond them and thus cannot cross this logical bridge into the geometric promised land of the *Elements*.

If weak students were likened to asses, Euclid himself met a similar fate at the hands of the Epicureans for his proof of proposition 20. To see why, we must describe a few of the intervening theorems of Book I.

After crossing the *pons asinorum*, Euclid showed how to bisect angles and construct perpendiculars with compass and straightedge and soon arrived at one of the critical theorems of Book I, commonly known as the exterior angle theorem. This result, which appeared as proposition 16, guaranteed that the exterior angle of any triangle exceeds either of the opposite and interior angles. That is (see Figure G.5), if we begin with △ *ABC* and extend side *BC* rightward to *D*, then both α and β are less than ∠*ACD*.

The exterior angle theorem was the first geometric inequality in the *Elements*. Where Euclid had previously shown sides or angles to be equal (as in the *pons*

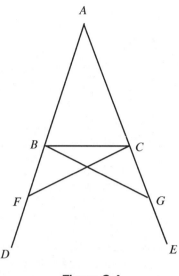

Figure G.4

asinorum), he here proved certain angles to be unequal. This theorem would play a significant role in the remainder of Book I.

This brings us to another inequality, proposition 19, whose diagram appears in Figure G.6. Euclid stated it as, "In any triangle the greater angle is subtended by the greater side," which in more modern notation becomes:

PROPOSITION 19: In $\triangle ABC$, if $\beta > \alpha$, then $\overline{AC} > \overline{BC}$ (i.e., $b > a$).

PROOF: Here we assume that $\beta > \alpha$. It is our job to prove that side AC opposite $\angle ABC$ is longer than side BC opposite $\angle BAC$.

Euclid separately considered the three possible cases: $b = a$, $b < a$, and $b > a$. His strategy was to show that the first two are impossible and thereby conclude that the third case must hold, as the theorem asserted. Such a technique is called a double *reductio ad absurdum*, or a proof by double contradiction. This powerful logical strategy was nowhere better employed than in Greek mathematics. Here is how Euclid handled it:

Case 1: Suppose $b = a$.

Referring to Figure G.6, we have $\overline{BC} = a = b = \overline{AC}$. This makes $\triangle ABC$ isosceles and so, invoking the *pons asinorum*, we conclude that the base angles are themselves congruent. That is, $\angle BAC = \angle ABC$, or equivalently $\alpha = \beta$. But this contradicts the initial assumption that $\beta > \alpha$. Hence we dismiss Case 1 as impossible.

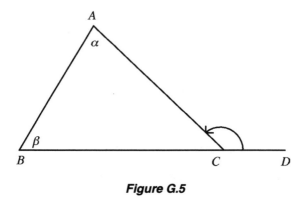

Figure G.5

Case 2: Suppose $b < a$.

Here we have the situation depicted in Figure G.7. Because AC is assumed to be shorter than BC, we can construct segment CD of length b where D falls within the longer side BC. Then draw AD to form $\triangle ADC$. This triangle, having two sides of length b, is isosceles and therefore has congruent base angles $\angle DAC$ and $\angle ADC$. But applying the exterior angle theorem to the narrow $\triangle ABD$, we deduce that

$\beta = $ interior $\angle ABD$

    $< $ exterior $\angle ADC$ by the exterior angle theorem

    $= \angle DAC$ because $\triangle DAC$ is isosceles

    $< \angle BAC$ because the whole is greater than the part

    $= \alpha$

In other words, $\beta < \alpha$, which contradicts the theorem's original stipulation that $\beta > \alpha$. Case 2, leading as it does to a contradiction, also bites the dust. We are left with:

Case 3: $b > a$.

This must be true because no alternatives remain, and the theorem is proved. ■

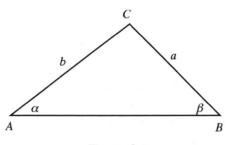

Figure G.6

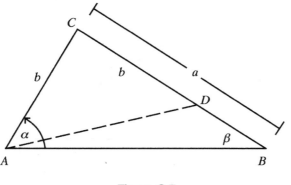

Figure G.7

We now have reached the proposition that so troubled the Epicurean philosophers. On the surface it sounds harmless enough:

PROPOSITION 20: In any triangle two sides taken together in any manner are greater than the remaining one.

Why the controversy? Why the derision? We quote the commentator Proclus:

> The Epicureans are wont to ridicule this theorem, saying it is evident even to an ass and needs no proof; it is as much the mark of an ignorant man, they say, to require persuasion of evident truths as to believe what is obscure without question. . . . That the present theorem is known to an ass they make out from the observations that, if straw is placed at one extremity of the sides, an ass in quest of provender will make his way along the one side and not by way of the two others.[8]

In short, even a dumb animal knows to take the straight route from C to B in Figure G.8 rather than go the long way around via A. So why, the Epicureans asked, did Euclid bother to prove something so blatantly obvious? Proclus gave an answer:

> It should be replied that, granting the theorem is evident to sense-perception, it is still not clear for scientific thought. Many things have this character; for example, that fire warms. This is clear to perception, but it is the task of science to find out how it warms.[9]

In the Euclidean spirit, the spirit that so typified Greek geometry, we must employ our faculties of reason to demonstrate that which an ass knows by instinct. Even a seemingly self-evident proposition cries out for a proof, and this Euclid was only too happy to provide. Building upon previous results, he reasoned as follows:

PROPOSITION 20: In $\triangle ABC$, $\overline{AC} + \overline{AB} > \overline{BC}$ (i.e., $b + c > a$)

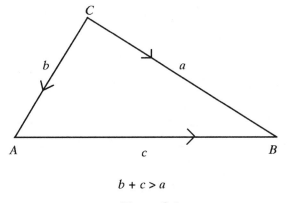

$$b + c > a$$

Figure G.8

PROOF: In Figure G.9, extend side BA to D so that $\overline{AD} = \overline{AC} = b$ and therefore $\overline{BD} = b + c$. This construction generates $\triangle DAC$, which is isosceles because it has two sides of length b. Considering the large $\triangle BDC$, we note that

$$\angle BCD > \angle ACD \qquad \text{because the whole is greater than the part}$$
$$= \angle BDC \qquad \text{because these are base angles of isosceles } \triangle DAC$$

So $\angle BCD$ is greater than $\angle BDC$. As Euclid had just proved that the greater side is opposite the greater angle, it follows that $\overline{BD} > \overline{BC}$; in other words, $b + c > a$, which is exactly what was to be proved. ■

This is a splendid little proof. It has its subtleties, its clever use of inequalities, its quiet elegance.

In Sir Arthur Conan Doyle's *A Study in Scarlet*, Dr. Watson described the deductive powers of Sherlock Holmes in these words: "His conclusions were as infallible as so many propositions of Euclid."[10] Watson was not alone in his high opinion of the Greek geometer. Centuries ago, the Arabic scholar al-Qifti said of Euclid, "nay, there was no one even of later date who did not walk in his footsteps,"[11] and the incomparable Albert Einstein added his own tribute: "If Euclid failed to kindle your youthful enthusiasm, then you were not born to be a scientific thinker."[12]

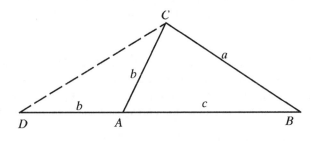

Figure G.9

What we have considered thus far, of course, is just the tip of the iceberg, just a sampler of what historian Morris Kline calls the Greeks' "grand exercise in logic."[13] We must leave them at this point. But in a sense, no mathematician can leave behind the legacy of the classical geometers. They gave demonstrative mathematics its start, honed its logical tools, and aimed it in directions it has traveled ever since. We end with the words of the twentieth-century British mathematician G. H. Hardy: "The Greeks . . . spoke a language which modern mathematicians can understand; as Littlewood said to me once, they are not clever schoolboys or 'scholarship candidates,' but 'Fellows of another college.' "[14]

There is no arguing with Hardy when he says, "Greek mathematics is the real thing."

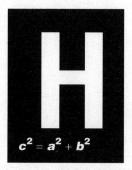

Hypotenuse

$$c^2 = a^2 + b^2$$

This chapter has a single object: to prove the Pythagorean theorem, a profound result about right triangles that has allowed mathematicians to cut corners for centuries. The theorem is surely among the greatest in all of mathematics. If our standard of greatness is the number of *different* proofs a theorem can boast, then Pythagoras' masterpiece wins hands down, for there are literally hundreds of arguments establishing its validity. An early twentieth-century professor named Elisha Scott Loomis collected and published 367 of them in a slightly kooky book called *The Pythagorean Proposition.*[1] Admittedly some of these proofs—which Loomis classifies as algebraic, geometric, dynamic, or quaternionic—are only minor variants of others, and *in toto* they tend to numb rather than illuminate, but their existence makes a point most clearly: This theorem has occupied mathematicians from classical times to the present.

We have neither the space nor the inclination to present hundreds of proofs of anything, but the Pythagorean theorem warrants at least a few. We shall consider three: one suggested by an ancient Chinese treatise, one popularized by seventeenth-century English mathematician John Wallis, and one discovered in 1876 by the U.S. politician and later president James A. Garfield. It is hoped that these will illustrate the agility of mathematicians in attacking the same problem from different angles.

First we are obliged to state the theorem. In modern form it says that, if ABC is a right triangle as shown in Figure H.1, then $c^2 = a^2 + b^2$, where a, b, and c are the lengths of the three sides. In our triangle, AC and BC are called the **legs**, and AB, the side opposite the right angle, is the *hypotenuse.*

As noted, this is the *modern* version. Those unfamiliar with Greek mathematics may be surprised to learn that it was regarded quite differently in classical times. The

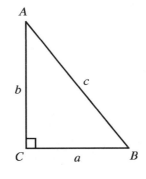

Figure H.1

Greeks had no algebraic symbolism, no formulas, no exponents. The equation $c^2 = a^2 + b^2$ would have been all Greek to them.

No, for the Greeks the Pythagorean theorem was a statement about *areas of squares*—literal, two-dimensional, four-side squares. Starting with right triangle *ABC*, they constructed squares upon the hypotenuse and legs as shown in Figure H.2. The theorem stated that the area of the square on the hypotenuse exactly equalled the sum of the areas of the squares on the legs. It was a remarkable and quite unexpected decomposition of one square's area into that of two smaller ones.

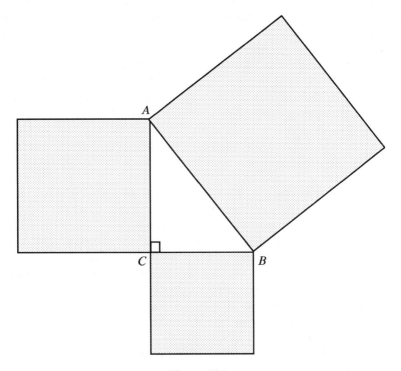

Figure H.2

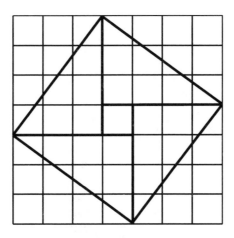

Figure H.3

Whether regarded algebraically or geometrically, the theorem is of supreme mathematical importance. But how is it proved? We begin with what is often called the Chinese proof.

The "Chinese Proof"

This is one of the most natural ways to do the job. In fact, many believe this was the way Pythagoras himself must have proved the result in the sixth century B.C. In all candor, there are those who doubt this is how Pythagoras did it, others who doubt that Pythagoras ever did it, and still others who doubt that Pythagoras even *existed*. Such are the problems of dealing with half-mythical characters from the distant past.

Although nothing survives of Pythagoras' work, the Chinese left behind a tangible indication of their reasoning. It appears in the *Chou pei suan ching*, a text dated anywhere from the time of Jesus to a thousand years earlier. Clearly the Chinese knew the theorem for right triangles with sides of length 3, 4, and 5, for they had a diagram called the *hsuan-thu* depicting a square askew within a square and shown in Figure H.3.[2]

It is true that this diagram was not accompanied by an axiomatic proof as Euclid would have given nor by a general argument addressing the Pythagorean relationship for *all* right triangles. We do not in reality have a proof at all. But the idea embodied in the *hsuan-thu* diagram can easily be fleshed out to establish the Pythagorean theorem.

The necessary prerequisites are few. One is the side-angle-side, or SAS, congruence scheme mentioned Chapter G. The other is the well-known theorem that the sum of the measures of a triangle's three angles is two right angles, or, in modern parlance, 180°. From this follows the trivial observation that the two acute angles of a *right* triangle must sum to 90°.

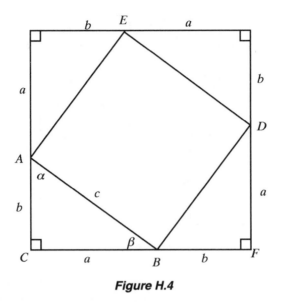

Figure H.4

With these modest underpinnings we begin. Letting $\triangle ABC$ be the right triangle shown in Figure H.4, mimic the Chinese diagram and construct a square with each side of length $a + b$ (by counting the blocks, one sees that for the 3-4-5 right triangle in Figure H.3, each side of the Chinese square was $3 + 4 = 7$ units long). Then draw BD, DE, and EA to create the tilted, four-sided figure within the large square. This is the generalized version of the *hsuan-thu*.

The big square has a right triangle in each corner, and these triangles, all having sides of length a and b surrounding right angles, are mutually congruent by SAS. Thus they are identical in all respects, so their four hypotenuses are equal—that is, $\overline{BD} = \overline{DE} = \overline{EA} = \overline{AB} = c$ —and their angles have measure α and β.

We now assert that quadrilateral $BDEA$ is a square. We have just observed that each of its four sides is of length c, so all that remains is to determine its angles. Consider, for instance, $\angle ABD$. Because CF is a straight line, we know that

$$180° = \angle CBA + \angle ABD + \angle DBF = \beta + \angle ABD + \alpha = 90° + \angle ABD$$

by our remark about the acute angles of a right triangle. Consequently

$$\angle ABD = 180° - 90° = 90°$$

and this corner of the inner figure is a right angle. The same argument applies to the other three corners as well, so $BDEA$, a quadrilateral with four equal sides and four right angles, is a square. As with any square, its area is just the product of its base and altitude: $c \times c = c^2$.

From here the conclusion follows easily. The large outer square of side $a + b$ has area of $(a + b)^2 = a^2 + 2ab + b^2$. But this outer square has been decomposed into five pieces—four congruent right triangles and a tilted inner square—so its area is also given by

$$4 \times \text{area}(\Delta \, ABC) + \text{area}(\text{square } BDEA) = 4\left(\frac{1}{2} \, ab\right) + c^2 = 2ab + c^2$$

We equate these two area expressions to get

$$a^2 + 2ab + b^2 = 2ab + c^2$$

and a subtraction of $2ab$ from each side yields the desired result:

$$a^2 + b^2 = c^2$$

The Chinese argument illustrates a gem of mathematical wisdom we met in the chapter on Euler: There is power in simultaneously approaching the same objective—in this case, the area of the large square—from two different directions. Such an approach can offer insights that no single viewpoint provides. Surely those who digest the Chinese proof will not soon again be hungry for knowledge.

A Similarity Proof

This proof of the Pythagorean theorem, attributed to the English mathematician John Wallis (1616–1703) but certainly much older, is regarded as one of the shortest and simplest of all. Superficially this assessment is accurate, for it takes very few lines to follow the argument from beginning to end. But the proof rests upon the notion of similar triangles, itself a concept requiring much spadework for a thorough development. Euclid waited until Book VI of the *Elements* to introduce similarity and thus could not have included a "Wallis-type" proof until then. As it was, he proved the Pythagorean theorem quite early, at the end of Book I. In this sense, *Euclid's* Pythagorean proof is shorter—closer to the postulates—than Wallis's. The true measure of the length of a proof must take into account not just the number of lines within the argument itself but the number of lines of mathematics that necessarily precede it.

It is also worth noting that, unlike the previous proof, this one does not approach the Pythagorean theorem by means of areas. Nowhere are squares broken up or reassembled. Instead the conclusion $c^2 = a^2 + b^2$ emerges as an algebraic consequence of results about lengths rather than as a geometric consequence of results about areas.

Nonetheless, this similarity proof is a good one. It requires us first to recall that two triangles are *similar* if the three angles of one are respectively congruent to the three angles of the other. Similarity is thus the perfect weapon for the compleat angler. In a less technical sense, similar triangles are those with the same shape but not necessarily the same size, so one triangle looks like an enlargement of the other. Clearly, similarity is a weaker condition than congruence, the latter requiring both

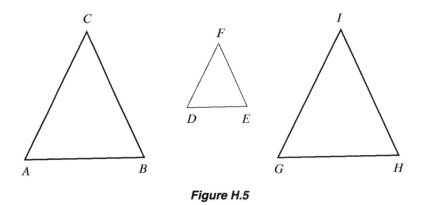

Figure H.5

the same shape *and* the same size. In Figure H.5, triangles *ABC* and *DEF* are similar whereas triangles *ABC* and *GHI* are congruent.

The critical feature of similar triangles is that their corresponding sides are proportional. For instance, if side *AB* is two-thirds as long as side *AC* in Figure H.5, then so, too, is side *DE* two-thirds as long as corresponding side *DF*. This proportionality of sides is tied up in what we mean by "same shape."

We now begin the similarity proof of the Pythagorean theorem. As before, let *ABC* be a right triangle with sides a and b, hypotenuse c, and with acute angles of measure α and β, where $\alpha + \beta = 90°$. From *C* draw *CD* perpendicular to *AB*, as shown in Figure H.6, and let $x = AD$.

Now consider the angles of $\triangle ADC$. One has measure α; one is a right angle; thus the remaining $\angle ACD$ is of size $180° - \alpha - 90° = 90° - \alpha = \beta$, because $\alpha + \beta = 90°$. It follows that $\triangle ADC$, having one angle of size α, one of size β, and one right angle, is similar to the original right triangle *ABC*. But the same can be said of $\triangle DCB$, because $\angle DCB = \angle ACB - \angle ACD = 90° - \beta = \alpha$. In short, the perpendicular *CD* has split right triangle *ABC* into two pieces, $\triangle ADC$ and $\triangle DCB$, that are smaller, similar copies of the original.

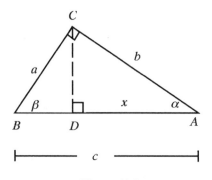

Figure H.6

We now invoke the proportionality of similar figures. Equating the ratios of hypotenuse to longer leg in $\triangle ADC$ and $\triangle ABC$, we conclude

$$\overline{AC}/\overline{AD} = \overline{AB}/\overline{AC}, \text{ or simply } \frac{b}{x} = \frac{c}{b}$$

from which it follows by cross-multiplication that $b^2 = cx$.

Next use the similarity of $\triangle CDB$ and $\triangle ABC$, along with the obvious fact that $\overline{DB} = \overline{AB} - \overline{AD} = c - x$, and equate the ratios of hypotenuse to shorter leg to get:

$$\overline{CB}/\overline{DB} = \overline{AB}/\overline{CB} \quad \text{or} \quad \frac{a}{c-x} = \frac{c}{a}$$

Cross-multiplication yields $a^2 = c(c - x) = c^2 - cx$.

Finally, add the results of these cross-multiplications and simplify:

$$a^2 + b^2 = (c^2 - cx) + cx = c^2$$

The Pythagorean theorem is proved once again, short and sweet.

Garfield's Trapezoidal Proof (1876)

U.S. presidents, whatever abilities they exhibit in other spheres, are seldom known for their mathematical powers. No professional mathematician has ever been elected to the White House, and recent presidents, oblivious to astronomical budget deficits, seem unable even to add properly.

Historically, however, some chief executives have possessed mathematical talent. One was George Washington, an accomplished surveyor, who endorsed mathematics with the following words:

> The investigation of mathematical truths accustoms the mind to method and correctness in reasoning, and is an employment peculiarly worthy of rational beings. . . . From the high ground of mathematical and philosophical demonstration, we are insensibly led to far nobler speculations and sublime meditations.[3]

Such statements tend to make Washington first in war, first in peace, and first in the hearts of mathematicians.

Abraham Lincoln was also a strong advocate of mathematics. As a young adult studying law, Abe recognized the need to sharpen his reasoning skills, to learn what it meant to prove a point by means of a sound logical argument. As he later recalled in an autobiographical sketch:

> I said, "Lincoln, you can never make a lawyer if you do not understand what demonstrate means"; and I left my situation in Springfield, went home to my father's house, and stayed there till I could give any proposition in the six books of Euclid at sight. I then found out what "demonstrate" means, and went back to my law studies."[4]

If Lincoln was telling the truth about mastering the 173 propositions in Books I–VI of the *Elements*—and who would accuse Honest Abe of lying?—this was no small accomplishment.

We should not overlook Ulysses S. Grant, who showed such mathematical promise while a cadet at the U.S. Military Academy at West Point that he dreamed of landing a faculty position. He later remembered his youthful vocational objectives: "My idea then was to get through the course, secure a detail for a few years as assistant professor of mathematics at the Academy, and afterwards obtain a permanent position as professor in some respectable college."[5] But, as Grant observed, "circumstances always did shape my course different from my plans." Of course, he ended up in the White House instead of the ivory tower.

Such achievements notwithstanding, none of these presidents was a mathematician. Thus, almost by default, the Mathematics Achievement Award for U.S. chief executives goes to James A. Garfield of Ohio, who published an original proof of the Pythagorean theorem in 1876.

Garfield was born near Cleveland in 1831 and divided his childhood between school and the odd jobs necessary to support his widowed mother. Young James, always a fine student, attended Western Reserve Academy and Hiram College in Ohio before transferring to Williams College in Massachusetts, from which he graduated in 1856. With his newly conferred degree, Garfield returned to teach mathematics at Hiram and seemed destined for the quiet academic life.

But those were not quiet days in the United States, for the nation stood on the brink of civil war. Amid heated discussions of secession and slavery, James Garfield was elected to the Ohio Senate in 1859. Radical in his politics and fiercely patriotic, he left academe to join the Union Army when war erupted in 1861. Interestingly, the math teacher turned out to be a fine soldier. Garfield quickly rose in the ranks until he was appointed chief of staff for Union General John Rosecrans.

In 1863, Garfield shifted from the U.S. Army to the U.S. House of Representatives, where he spent the next 17 years as a Radical Republican intent upon reforming, if not punishing, the South. It was during this time that Congressman Garfield discovered his Pythagorean proof while "in some mathematical amusements and discussions with other M.C.s"[6] and published it in the *New England Journal of Education,* a periodical devoted to "education, science, and literature."

In 1880 James A. Garfield earned the Republican nomination for the presidency and narrowly defeated another Civil War hero, the Democrat Winfield Scott Hancock, in that fall's election. At his inauguration in March of 1881, our mathematical president promised to improve the educational opportunities of all Americans, because, "It is the high privilege and sacred duty of those now living to educate their successors and fit them, by intelligence and virtue, for the inheritance which awaits them."[7]

But promise was about all that came from the Garfield administration, for on July 2, 1881, having served less than four months, he was shot by a disgruntled office seeker while boarding a train in Washington. Although Garfield's life proba-

James A. Garfield
(Courtesy of Muhlenberg College Library)

bly could be saved with today's medical techniques, the wound eventually proved fatal to the president, who lingered until mid-September before death overtook him.

In its grief the nation named towns, streets, schools, and children after its fallen leader. An impressive tomb was built in Cleveland and thousands of visitors paid their respects. Politically, his was a life with its greatest dreams left unfulfilled. But he left a mark upon mathematics.

To understand Garfield's proof, two preliminaries are necessary. One is the well known angle-side-angle, or ASA, congruence scheme, which says that if triangles have two angles and the included side of one respectively equal to two angles and the included side of the other, the triangles are congruent. The other is a formula for the area of a trapezoid. A *trapezoid*, of course, is a four-sided figure having a pair of

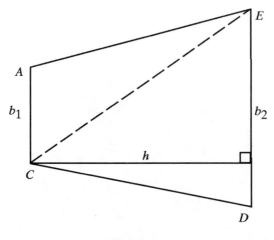

Figure H.7

opposite sides parallel. Finding its area is not difficult, because a diagonal splits the trapezoid into a pair of triangles with the same height.

Thus, in Figure H.7 we have trapezoid $ACDE$ with (vertical) parallel sides AC and DE of lengths b_1 and b_2, respectively, and with height h, the perpendicular distance between the parallel sides. Diagonal CE divides the figure into the two triangles, and consequently

$$\text{area(trapezoid)} = \text{area}(\triangle ACE) + \text{area}(\triangle CED)$$

$$= \frac{1}{2}\, b_1 h + \frac{1}{2}\, b_2 h = \frac{1}{2}\, h(b_1 + b_2)$$

In other words, the area of a trapezoid is half the product of its height and the sum of its bases.

We now consider Garfield's proof (although we have reoriented and relabeled his diagram in Figure H.8). As always, start with right triangle ABC having right angle at C, legs a and b, and hypotenuse c. From B construct BE perpendicular to AB with $\overline{BE} = c$ and from E draw ED perpendicularly downward, where D is the intersection of the foot of the perpendicular and the rightward extension of line CB. Finally, draw AE.

Following Garfield, we examine the consequences of these constructions. First, it is clear that

$$\angle DBE = 180° - \angle ABE - \angle CBA = 180° - 90° - \beta = 90° - \beta = \alpha$$

since $\alpha + \beta = 90°$. Because $\angle DBE = \alpha$ and $\angle BDE$ is right, it follows that $\angle BED = \beta$. Thus $\triangle BED$ is congruent to $\triangle ABC$ by the ASA congruence scheme, where the equal sides are BE and AB. From congruence we conclude the corresponding sides are equal: $\overline{BD} = \overline{AC} = b$ and $\overline{DE} = \overline{BC} = a$.

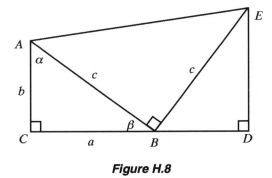

Figure H.8

Further, quadrilateral *ACDE* is a trapezoid because its opposite sides *AC* and *DE*, both perpendicular to *CD*, are parallel. Therefore—and here Garfield's insight becomes apparent—we can determine the area of the trapezoid *ACDE* in two different ways. From the trapezoidal area formula above, we know that

$$\text{area(trapezoid } ACDE) = \frac{1}{2}\, h(b_1 + b_2) = \frac{1}{2}\,(b + a)\,(b + a)$$

because the parallel bases are of length $b_1 = \overline{AC} = b$ and $b_2 = \overline{DE} = a$, and the distance between the parallel sides is $h = \overline{CD} = \overline{BD} + \overline{BC} = b + a$.

On the other hand, the area of trapezoid *ACDE* is the sum of the areas of the three right triangles of which it is composed:

$$\text{area(trapezoid } ACDE) = \text{area}(\triangle\, ACB) + \text{area}(\triangle\, ABE) + \text{area}(\triangle\, BDE)$$

$$= \frac{1}{2}\, ab + \frac{1}{2}\, c^2 + \frac{1}{2}\, ab = ab + \frac{1}{2}\, c^2$$

Finally, equate these two expressions for the trapezoid's area and mop up algebraically:

$$\frac{1}{2}\,(b + a)\,(b + a) = ab + \frac{1}{2}\, c^2 \rightarrow \frac{1}{2}\,(b^2 + 2ab + a^2) = ab + \frac{1}{2}\, c^2$$

Upon doubling each side we get $b^2 + 2ab + a^2 = 2ab + c^2$ and a subtraction of $2ab$ gives the desired result:

$$a^2 + b^2 = c^2$$

Garfield's is really a very clever proof. Here again we see the benefits of looking at the trapezoid's area from two different viewpoints. As the author of the *New England Journal* article wryly observed, "we think it something on which the members of both houses can unite without distinction of party."[8]

Yet there is something vaguely familiar about Garfield's diagram. The reader may note that, if we augment Garfield's figure by including its mirror image across

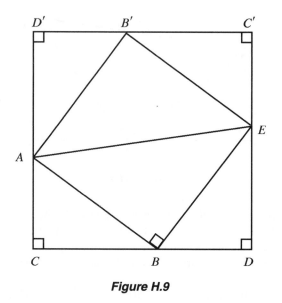

Figure H.9

segment *AE*, we find ourselves staring at the *hsuan-thu* of the Chinese proof (see Figure H.9). Garfield had stumbled upon a variation of the ancient recipe.

We have thus given three proofs of the theorem of Pythagoras, enough (we hope) to convince even the most hardened skeptic. Of course, one might question the necessity of proving the same result in multiple ways. Aren't these extra proofs redundant?

In a practical sense they are. There is no *logical* necessity to establish a theorem more than once. But there is an aesthetic motivation to re-explore the same ground. Just because someone once wrote a love song should not prevent other songwriters from doing the same, albeit with a different melody, an altered lyric, a modified rhythm. Likewise, these various proofs of the Pythagorean theorem reveal different mathematical melodies and rhythms, no less beautiful because they address an old theme.

It is perhaps fitting to add a brief word about the Pythagorean theorem's converse. *Converse* is a term from logic with a precise meaning. If we begin with the statement, "If A, then B," and interchange hypothesis and conclusion, we get the related statement, "If B, then A." The latter is called the **converse** of the original; the original's hypothesis is the converse's conclusion and vice versa.

A moment's thought reveals that a proposition may be true and its converse be true as well. For instance, the statement, "If a triangle has three congruent sides, then it has three congruent angles," and its converse, "If a triangle has three congruent angles, then it has three congruent sides," are both valid geometric theorems.

On the other hand, a statement may be true yet its converse be false. The proposition, "If Rex is a dog, then Rex is a mammal," is valid, as any zoologist will confirm.

But its converse, "If Rex is a mammal, then Rex is a dog," is untrue, as my Uncle Rex is quick to point out.

The converse of the Pythagorean theorem is:

If $c^2 = a^2 + b^2$, then $\triangle ABC$ is a right triangle.

This converse is true, as Euclid proved in the final proposition of the first book of the *Elements*. His argument was another splendid example of Greek geometry in action. It established that a triangle is right *if and only if* the square on the hypotenuse is the sum of the squares on the other two sides. This gives a complete characterization of right triangles, and geometers cannot ask for more.

With this our discussion of the Pythagorean theorem comes to an end (although we stress that those with heartier appetites will find 364 *more* proofs in Loomis's book). But even a flood of different proofs will not dilute the significance of this great result. For, no matter how often it is proved, the theorem of Pythagoras always manages to retain its beauty, its freshness, and its eternal sense of wonder.

Isoperimetric Problem

circle = max

According to classical mythology, Princess Dido, whose murderous brother Pygmalion was king of Tyre, fled her homeland, sailed the Mediterranean with a band of compatriots, and landed on the northern coast of Africa. In the *Aeneid*, Virgil tells us:

> Here they bought ground; they used to call it Byrsa,
> That being a word for bull's hide; they bought only
> What a bull's hide could cover.[1]

That is, Dido's acquisition of land for a great new city was to be limited to the area she could enclose with the hide of a bull.

Being clever, Dido first cut the hide into a number of long, thin strips. Being even more clever, she placed these rawhide strands in the shape of a great semicircle whose diameter ran along the seashore. Within this large area was built the city of Carthage.

In this fanciful story appear the mythical origins of two perfectly real phenomena. One is the founding of Carthage, a city-state that exercised tremendous power over the Mediterranean world, tangled with the equally powerful Romans in three Punic Wars between 264 and 146 B.C., and gave military historians the specter of Hannibal's elephants crossing the Alps in an improbable flank attack upon Italy. Ruins along the Tunisian coast are all that now remain of Carthage, a sad reminder of Rome's treatment of its enemies.

But the story of Dido also provides the mythic origins of a famous mathematical question. How should a fixed perimeter (the bullhide strips) be configured so as to enclose the maximum area along the seashore? Dido reasoned that a semicircle

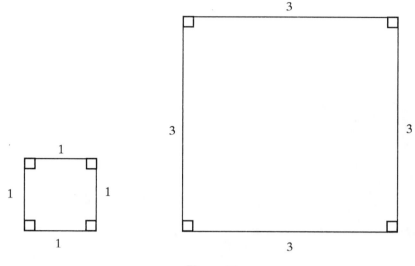

Figure I.1

would do the job and in the process bequeathed to us what is sometimes called Dido's problem.

Of course, mathematicians feel uncomfortable attributing results to mythological beings, and they hate problems requiring bullhide. So it is customary today to call this the isoperimetric problem (*iso* = same, *perimetric* = boundary) and state it formally as:

> To determine, from among all curves of the same perimeter,
> the one enclosing the largest area.

This is a wonderful challenge. It gave the Greek geometers a good workout and resurfaced 2,000 years later as a test for the newly developed subject of calculus.

Enclosing different areas within the same perimeter may seem paradoxical. Proclus, whom we met in Chapter G defending Euclid against the ridicule of the Epicureans, observed that many of his contemporaries were ill-informed about this matter. They thought, said Proclus, that the larger the perimeter of a figure, the larger the area within. This, of course, is *sometimes* the case, as with the squares of Figure I.1: The square on the left has perimeter 4 and area of 1 and its right-hand counterpart has larger perimeter (12) and larger area (9).

But such a relationship is not necessary, as Proclus observed. The parallelogram on the left in Figure I.2 has been formed by adjoining two 3–4–5 right triangles each of area $1/2bh = 1/2(4 \times 3) = 6$. Its perimeter is 18 and its area is $6 + 6 = 12$. The 4×4 square at the right has smaller perimeter (16) but larger area (16).

So we dare not compare areas by comparing perimeters, a point Proclus drove home when he mocked "geographers who infer the size of a city from the length of its walls" or when he described unscrupulous land speculators who exchanged lands

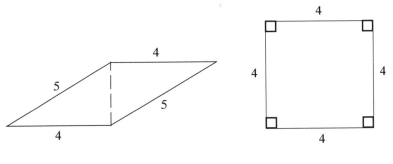

Figure I.2

of longer periphery (but smaller area) for those of shorter periphery (but larger area) and yet "gained a reputation for superior honesty."[2]

Enlarging the perimeter of a figure need not increase the area within. But what if we are given a *fixed* perimeter to work with? What can be said about the area it surrounds?

To put this in concrete terms, imagine that we have a specified length of rope—say, 600 feet—that we wish to configure in such a way as to maximize the area it encloses. Obviously there are many different areas that can be contained by this allotment of rope. A long, narrow rectangle with dimensions 1×299 has a perimeter of 600 feet and an area of 299 square feet, whereas a fatter rectangle of 100×200 has the same 600-foot perimeter but now surrounds a much larger 20,000 square feet of area (see Figure I.3).

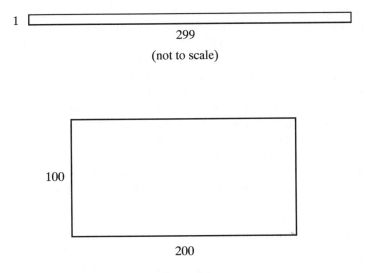

Figure I.3

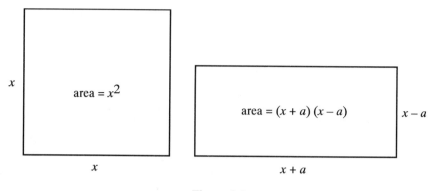

Figure I.4

Of all rectangles with the same perimeter, it is the *square* that encloses the greatest area. This is easy to prove using the maximizing techniques of differential calculus from Chapter D, but we shall give a more elementary argument.

Suppose we have a fixed perimeter from which we make a square of side x. The enclosed area, shown in Figure I.4, is obviously x^2. If we transform the square into a rectangle by lengthening the horizontal sides to $x + a$, we must simultaneously shrink the vertical sides to $x - a$ in order to maintain the same perimeter. Consequently, the rectangle's area will be

$$(x + a)(x - a) = x^2 - a^2$$

which is clearly less than x^2. In other words, a nonsquare rectangle of a given perimeter encloses an area smaller by a^2 than a square of the same perimeter.

This principle was proved by the Greek mathematician Zenodorus using purely geometric ideas sometime around 200 B.C. None of his original works survive, so all we know of him is what can be gleaned from passing references of other writers. These later commentators recorded that Zenodorus wrote a treatise called *On Isoperimetric Figures* in which many key results appeared.

For instance, Zenodorus proved that of all polygons having the same number of sides, it is the *regular* polygon (as introduced in Chapter C) that encloses the greatest area.[3] Consequently an equilateral triangle contains a greater area than any other triangle of the same perimeter, and a square is more spacious than any quadrilateral of the same perimeter. The proof of this general theorem is by no means simple.

But the isoperimetric challenge is not restricted to triangles or rectangles. In fact, if we bend our 600 feet of rope into a regular hexagon with each of its six sides 100 feet long, it is even roomier than the square (see Figure I.5). A proof of this fact follows:

Let O be the center of the regular hexagon of side 100 feet and let h be the length of the line from O perpendicular to a side of the hexagon (as we mentioned in Chapter C, h is called the apothem of the regular polygon). Some elementary geometry reveals that the apothem is perpendicular to and bisects side AB, so that $\overline{AD} = 50$,

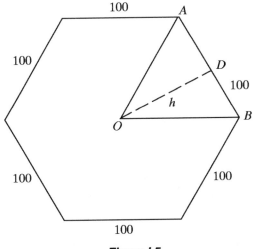

Figure I.5

and $\overline{OA} = \overline{OB} = \overline{AB} = 100$ feet. Applying the Pythagorean theorem to right triangle *ODA*, we see that $100^2 = 50^2 + h^2$, and so

$$h = \sqrt{100^2 - 50^2} = \sqrt{7500} \text{ feet}$$

It follows that Δ *OAB* has area

$$\frac{1}{2} bh = \frac{1}{2} (100)\sqrt{7500} = 50\sqrt{7500}$$

and the entire regular hexagon has area six times as great, namely $300\sqrt{7500} \approx$ 25980.76 square feet. This exceeds the 20,000-square-foot area of the square with the same perimeter considered above.

If, however, our 600 feet of rope is configured into a regular octagon (with each side 75 feet long), it encloses 27,159.90 square feet; and a regular dodecagon (with each of its 12 sides 50 feet long) encloses 27,990.38 square feet (see Figure I.6).

This chain of examples suggests that if the perimeter is unchanged but the number of sides of the regular polygon increases, then the enclosed area simultaneously increases. According to later commentators, Zenodorus stated this principle as

> Of all rectilinear figures having an equal perimeter—I mean equilateral and equiangular figures—the greatest is that which has most angles[4]

and provided a proof.

At this point we indulge in a fairly extensive digression by jumping ahead many centuries to the late classical mathematician Pappus, who flourished around A.D. 300. Pappus wrote a treatise describing Zenodorus' work and gave an example of the previous isoperimetric principle in action. It appeared, strangely enough, when Pappus interrupted his scholarly work to discuss the mathematical aptitude of *bees*,

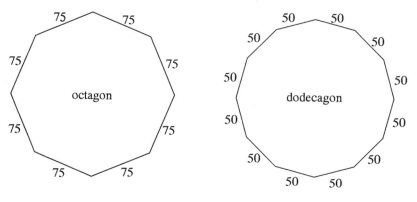

Figure I.6

insects he obviously held in the highest regard. In what must be one of the most anthropomorphic statements ever made, Pappus asserted that bees "Believ[e] themselves, no doubt, to be entrusted with the task of bringing from the gods to the more cultured part of mankind a share of ambrosia."[5] Having thus suggested that bees make honey primarily for human consumption, Pappus noted that they naturally would want to store it without waste, by depositing it in cells arranged so that "nothing else might fall into the interstices" and thus be lost. The cells of the honeycomb, in other words, must be constructed so as to leave no gaps.

Assuming that the honeycomb is made of *identical regular polygons* (we might as well join Pappus in believing that bees imposed such a requirement), we prove the following.

PROPOSITION: There are only three ways to arrange identical regular polygons about a common vertex without "interstices."

PROOF: As a first step toward verifying this assertion, we determine the number of degrees in each angle of a regular polygon having 3, 4, or, in general, *n* sides.

Fortunately, this is not difficult. Suppose we have a regular polygon of *n* sides each of whose angles is of size α. (Of course, $n \geq 3$, because a polygon cannot have two or fewer sides.) From its center O we draw lines to the vertices, thereby subdividing the regular polygon into *n* identical triangles as depicted in Figure I.7. The trick now is to calculate in two different ways the degree total of these triangles.

On the one hand, because there are *n* triangles each containing 180°, the total number of degrees of the triangles within the polygon is obviously $n \times 180°$. But look again at the angle sum of these *n* triangles. Their vertices all meet at point O, so the sum of their vertex angles is just the number of degrees obtained by making a complete revolution around that point—that is, 360°. Meanwhile, their base angles, of which there are a total of 2*n*, each have degree measure $\alpha/2$. Therefore, the total number of degrees in the triangles composing the polygon is $360° + 2n(\alpha/2) =$

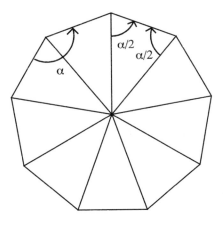

Figure I.7

$360° + n\alpha$. Equate these two different expressions for the total angular measure of the triangles and solve for α:

$$360° + n\alpha = n \times 180°$$

$$n\alpha = n \times 180° - 360°$$

$$\alpha = \frac{n \times 180° - 360°}{n} = 180° - \frac{360°}{n}$$

This formula gives the number of degrees in each internal angle of a regular n-gon.

We apply the formula to a few specific cases. If $n = 3$, each angle in a regular (i.e., equilateral) triangle contains

$$\alpha = 180° - \frac{360°}{3} = 180° - 120° = 60°$$

as is well known. Each angle of a square ($n = 4$) contains $180° - 360°/4 = 180° - 90° = 90°$, a right angle; each angle of a regular pentagon ($n = 5$) contains $180° - 360°/5 = 180° - 72° = 108°$; and each angle of a regular hexagon ($n = 6$) contains $180° - 360°/6 = 120°$.

Fine. But the proposition under consideration goes further, for we are trying to arrange regular polygons about a common vertex so as to leave no spaces. This is the sort of configuration displayed by floor tiles, which fit together perfectly and thus keep spilled milk from dripping down into the basement.

We must determine *how many* regular polygons can meet at each vertex if gaps are to be avoided. We therefore let k be the number of identical regular polygons meeting at a vertex, as shown in Figure I.8. It is clear that $k \geq 3$, for we cannot have two or fewer polygons meeting about a vertex.

Denote by n the number of sides of each regular polygon. We have just determined that each polygonal angle contains $180° - 360°/n$, and because k of them

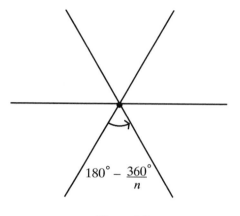

$180° - \dfrac{360°}{n}$

Figure I.8

meet at a vertex, we see that the total number of degrees around the vertex is $k \times (180° - 360°/n)$. But the total number of degrees around each vertex is also 360°. Equating these two expressions yields

$$360° = k \times \left(180° - \frac{360°}{n}\right)$$

from which it follows, upon dividing both sides by 360°, that

$$1 = k\left(\frac{1}{2} - \frac{1}{n}\right)$$

Finally, because we know that $k \geq 3$, we generate the critical inequality

$$1 = k\left(\frac{1}{2} - \frac{1}{n}\right) \geq 3\left(\frac{1}{2} - \frac{1}{n}\right) = \frac{3}{2} - \frac{3}{n}$$

Hence

$$\frac{3}{n} \geq \frac{3}{2} - 1 = \frac{1}{2}$$

Cross-multiply to conclude that $3 \times 2 \geq n \times 1$, or simply $n \leq 6$.

This inequality restricts the kinds of regular polygons that can be arranged about a common vertex, for it shows that each must have six or fewer sides. We investigate the possible cases separately, with illustrations in Figure I.9:

a. If $n = 3$, each polygon is an equilateral triangle with 60° per angle. We then can put $360°/60° = 6$ equilateral triangles together at each vertex without gaps. This is thus a possible configuration for floor tiles or honeycombs.

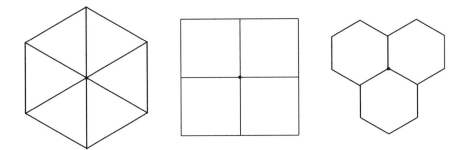

Figure I.9

b. If $n = 4$, each polygon is a square with 90° per angle. We can clearly assemble 360°/90° = 4 squares at each vertex. Again, bees and linoleum installers take note.

c. For $n = 5$, we have seen that each angle of a regular pentagon contains 108°. But 108° does not go evenly into 360°, as 360°/108° = 3 1/3. Thus, regular pentagons cannot fill all space about a point without leaving gaps. This case must be discarded.

d. For $n = 6$, we have a regular hexagon. Each interior angle contains 120° and so we can assemble 360°/120° = 3 hexagons at each vertex.

Because $3 \leq n \leq 6$, there are no other possibilities. Therefore, as claimed, the only arrangements of identical regular polygons that avoid gaps are six equilateral triangles, four squares, or three regular hexagons. ■

Whew! This proof was quite an accomplishment for Pappus' bees and must have kept their little antennae quivering for weeks. But, having displayed so much mathematical acumen, these bugs faced one final question: Which of the three possible arrangements is best for their hives?

Here, at last, they exhibited a keen understanding of the isoperimetric principle: To get the maximum honey storage (i.e., the maximum cross-sectional area) for the same amount of honeycomb (i.e., the same perimeter), they choose the polygon with the most sides—the regular hexagon! Sure enough, as any entomologist will confirm, bees make hexagonal combs. Pappus wrote, "Bees, then, know just this fact which is useful to them, that the hexagon is greater than the square and the triangle and will hold more honey for the same expenditure of material in constructing each."[6] Here Pappus attributed to bees a mathematical aptitude that would exceed the graduation requirements of most modern universities. To him, bees were miniature geometers with whom he could see eye to eye to eye.

Zenodorus' isoperimetric principle—for a fixed perimeter, the more sides of a regular polygon, the greater the enclosed area—led to an immediate and remarkable corollary: A circle, the limiting figure of regular polygons with ever-increasing num-

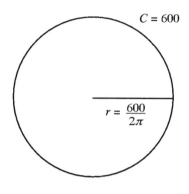

Figure I.10

bers of sides, encloses greater area than any regular polygon of equal perimeter. This, too, Zenodorus is said to have proved.

There is something wonderfully Greek about this conclusion. Recall that even before the time of Euclid the Greeks had enshrined the straight line and the circle as the two indispensable geometric figures, the two shapes constructible with geometric tools. Of course, the straight line is the shortest distance between two points. And Zenodorus showed that the circle was the greatest encloser of area. The figure swept out by Euclid's compass enclosed the maximum area possible for the given perimeter. Is this not further indication of the ideal form the Greeks so admired?

To illustrate the isoperimetric principle, we return again to our 600 feet of rope and bend it into the circle shown in Figure I.10. Because the circumference is $2\pi r$, we find that $600 = 2\pi r$, which implies $r = 600/2\pi$. As shown in Chapter C, circular area is given by $A = \pi r^2$, so the area of this circle is

$$A = \pi \left(\frac{600}{2\pi}\right)^2 = \frac{360,000}{4\pi} = \frac{90,000}{\pi} \approx 28,647.89 \text{ square feet}$$

greater than even the regular dodecagon considered above. As Zenodorus observed, the circle is larger than any regular polygon with the same perimeter.

Why, then, did Dido place her strips of bullhide in a semicircular arrangement? The answer is that she was using the shoreline as one boundary of her property and thus did not need to expend bullhide for that part. Under that modification the semicircle is indeed the best. Had Carthage been founded not along the seacoast but rather inland—say, somewhere in Kansas—then Dido certainly would have enclosed her property in a circle.

But even these remarkable theorems did not fully resolve the isoperimetric problem, for it is conceivable that we might exceed a circle's area by assembling not regular polygons (which Zenodorus considered) but parabolas, ellipses, or some other irregular curves (which he did not). Because of the generality of the curves involved, the final solution eluded the Greek mathematicians. As noted in Chapter B, it was over an isoperimetric challenge that Johann and Jakob Bernoulli engaged in one of their typical squabbles and thereby helped create the truly sophisticated branch of mathematics now called the calculus of variations.

But even for the more general isoperimetric problem, it turns out that the ancient answer was still the correct one. Of all curves with fixed perimeter—polygons, ellipses, parabolas, etc.—the trusty circle will enclose the maximum area. It is an amazing property.

Down through the centuries, there have been a handful of seminal problems that have intrigued, and frustrated, mathematicians. Some, such as the primality of Mersenne numbers from Chapter A, remain open. Others, such as the trisection of angles with compass and straightedge, were resolved only after centuries of effort, as we see in Chapter T. The isoperimetric problem—so simple to state yet so difficult to prove—holds a similar distinction. With contributions from Dido and Zenodorus, from the Bernoullis and the bees, it has surely earned its place among the mathematical classics.

ustification

Q.E.D.

"Proof," says mathematician Michael Atiyah, "is the glue that holds mathematics together."[1] Such a view certainly suggests that proving—or justifying—is the embodiment of mathematics.

This point can be debated. Mathematics is a discipline vast enough to include such activities as estimating quantities, constructing counterexamples, testing special cases, and solving utterly routine problems. A mathematician is not necessarily engaged 24 hours a day in proving theorems.

But if logical justification of theoretical propositions is not the only activity of mathematics, it is surely the discipline's trademark. Nothing sets mathematics apart from other scholarly pursuits like its reliance on proofs, reason, and logical deduction. In comparing mathematics and logic, Bertrand Russell (1872–1970) asserted, "it has become wholly impossible to draw a line between the two; in fact, the two are one."[2]

Already this book has examined a number of mathematical justifications. In the first chapter we proved the infinitude of primes, and in Chapter H we proved and reproved the theorem of Pythagoras. As mathematical arguments go, these were fairly simple. Other proofs require many pages, many chapters, or even many volumes to reach their conclusions. The associated demands upon the intellect are not to everyone's taste, as was suggested when a self-deprecating Charles Darwin wrote, "My power to follow a long and purely abstract train of thought is very limited; and therefore I could never have succeeded with metaphysics or mathematics."[3] Or, in the more succinct words of John Locke, "Mathematical proofs, like diamonds, are hard as well as clear."[4]

What exactly is a proof of a mathematical theorem? This question is not as open and shut as it might seem, for it involves philosophical and psychological as well as

362   PROLEGOMENA TO CARDINAL ARITHMETIC   [PART II

$*54\cdot42$. $\vdash :: \alpha \,\epsilon\, 2 . \supset :. \beta \,C\, \alpha . \underset{\exists}{\lrcorner}\,!\,\beta . \beta \,\text{\textdbend}\, \alpha . \equiv . \beta \,\epsilon\, \iota``\alpha$

Dem.

$\vdash . *54\cdot4 . \quad \supset \vdash :: \alpha = \iota`x \cup \iota`y . \supset :.$

$\qquad\qquad \beta \,C\, \alpha . \underset{\exists}{\lrcorner}\,!\,\beta . \equiv : \beta = \Lambda . \text{v} . \beta = \iota`x . \text{v} . \beta = \iota`y . \text{v} . \beta = \alpha : \underset{\exists}{\lrcorner}\,!\,\beta :$

$[*24\cdot53\cdot56 . *51\cdot161] \qquad \equiv : \beta = \iota`x . \text{v} . \beta = \iota`y . \text{v} . \beta = \alpha \qquad\qquad (1)$

$\vdash . *54\cdot25 . \text{Transp.} *52\cdot22 . \supset \vdash : x \,\text{\textdbend}\, y . \supset . \iota`x \cup \iota`y \,\text{\textdbend}\, \iota`x . \iota`x \cup \iota`y \,\text{\textdbend}\, \iota`y :$

$[*13\cdot12] \qquad\qquad \supset \vdash : \alpha = \iota`x \cup \iota`y . x \,\text{\textdbend}\, y . \supset . \alpha \,\text{\textdbend}\, \iota`x . \alpha \,\text{\textdbend}\, \iota`y \qquad\qquad (2)$

$\vdash . (1) . (2) . \supset \vdash :: \alpha = \iota`x \cup \iota`y . x \,\text{\textdbend}\, y . \supset :.$

$\qquad\qquad \beta \,C\, \alpha . \underset{\exists}{\lrcorner}\,!\,\beta . \beta \,\text{\textdbend}\, \alpha . \equiv : \beta = \iota`x . \text{v} . \beta = \iota`y :$

$[*51\cdot235] \qquad\qquad\qquad\qquad\qquad \equiv : (\underset{\exists}{\lrcorner}z) . z \,\epsilon\, \alpha . \beta = \iota`z :$

$[*37\cdot6] \qquad\qquad\qquad\qquad\qquad \equiv : \beta \,\epsilon\, \iota``\alpha \qquad\qquad (3)$

$\vdash . (3) . *11\cdot11\cdot35 . *54\cdot101 . \supset \vdash . \text{Prop}$

$*54\cdot43$. $\vdash :. \alpha, \beta \,\epsilon\, 1 . \supset : \alpha \cap \beta = \Lambda . \equiv . \alpha \cup \beta \,\epsilon\, 2$

Dem.

$\vdash . *54\cdot26 . \supset \vdash :. \alpha = \iota`x . \beta = \iota`y . \supset : \alpha \cup \beta \,\epsilon\, 2 . \equiv . x \,\text{\textdbend}\, y .$

$[*51\cdot231] \qquad\qquad\qquad\qquad\qquad \equiv . \iota`x \cap \iota`y = \Lambda .$

$[*13\cdot12] \qquad\qquad\qquad\qquad\qquad \equiv . \alpha \cap \beta = \Lambda \qquad\qquad (1)$

$\vdash . (1) . *11\cdot11\cdot35 . \supset$

$\qquad \vdash :. (\underset{\exists}{\lrcorner}x, y) . \alpha = \iota`x . \beta = \iota`y . \supset : \alpha \cup \beta \,\epsilon\, 2 . \equiv . \alpha \cap \beta = \Lambda \qquad (2)$

$\vdash . (2) . *11\cdot54 . *52\cdot1 . \supset \vdash . \text{Prop}$

From this proposition it will follow, when arithmetical addition has been defined, that $1 + 1 = 2$.

Figure J.1

Russell and Whitehead prove 1 + 1 = 2

(From *Principia Mathematica, Vol. 1* by Alfred North Whitehead and Bertrand Russell, 1935. Reprinted with the permission of Cambridge University Press.)

mathematical issues. Aristotle sensed as much when he described proof as "a matter not of external discourse but of meditation within the soul."[5]

Equally valid is Russell's observation that the mathematician can never put onto paper the "complete process of reasoning," but rather must settle for "such an abstract of the proof as is sufficient to convince a properly instructed mind."[6] He was suggesting that any mathematical statement is predicated upon other statements and definitions, which are predicated upon still more statements and definitions, so that it would be foolhardy to require proofs to track back through *every* logical preliminary.

Russell seemed to ignore his own advice when he and Alfred North Whitehead (1861–1947) produced the magnificently tedious work *Principia Mathematica* during the early years of the twentieth century. In it they tried to trace all of mathematics back to fundamental logical principles and spared no detail in the process. The result was excruciating. So deliberate was their development that it took until page 362, in section 54.43 of a chapter titled "Prolegomena to Cardinal Arithmetic," before they could at last prove that $1 + 1 = 2$ (see Figure J.1). The *Principia* was justification gone berserk.

In this chapter we try to stay sane. For our purposes, a proof is any argument, carefully crafted within the rules of logic, that is unassailably convincing as to the validity of an assertion. Questions such as, "Convincing to whom?" or, "Unassailable by whose standards?" are left for another time.

Of course, we might alternately consider what a proof is not. It is not justification by appeals to intuition or to common sense or, worse yet, to intimidation. Neither are we talking about proof "beyond a shadow of a doubt" as with proof of guilt in a criminal proceeding. Mathematicians like to think of proof as being not merely beyond reasonable doubt but beyond *all* doubt.

A discussion of mathematical justification can lead in many different directions. Here we present four important maxims and raise one increasingly significant question regarding the nature of proof in mathematics.

Maxim #1: A Few Cases Aren't Enough

In the sciences, and certainly in everyday life, we tend to accept the truth of a principle when experiments repeatedly confirm it. If the confirming examples are numerous enough, we say we have a "proven law."

But to a mathematician the outcomes of a few cases, though perhaps suggestive, are by no means a proof. As an example of this phenomenon, consider

CONJECTURE: Upon substituting a positive integer into the polynomial $f(n) = n^7 - 28n^6 + 322n^5 - 1,960n^4 + 6,769n^3 - 13,132n^2 + 13,069n - 5,040$, we always get the same positive integer back. Symbolically the assertion states that $f(n) = n$ for any whole number n.

Is it true? An obvious and perfectly reasonable starting point is to try a few numbers and see what happens. For $n = 1$ we get

$$f(1) = 1 - 28 + 322 - 1,960 + 6,769 - 13,132 + 13,069 - 5,040 = 1$$

as claimed. If we substitute $n = 2$, the outcome is

$$f(2) = 2^7 - 28(2^6) + 322(2^5) - 1,960(2^4) + 6,769(2^3)$$
$$- 13,132(2^2) + 13,069(2) - 5,040 = 2$$

which again is supportive. The reader is invited to take out a calculator and verify that $f(3) = 3, f(4) = 4, f(5) = 5, f(6) = 6$, and even

$$f(7) = 7^7 - 28(7^6) + 322(7^5) - 1,960(7^4) + 6,769(7^3)$$
$$- 13,132(7^2) + 13,069(7) - 5,040 = 7$$

Evidence on behalf of the claim seems to be building. Some people, especially if unenthusiastic about performing mindless calculations, may be ready to pronounce the statement true.

But it isn't. Substituting $n = 8$ yields

$$f(8) = 8^7 - 28(8^6) + 322(8^5) - 1,960(8^4) + 6,769(8^3)$$
$$- 13,132(8^2) + 13,069(8) - 5,040 = 5,048$$

rather than the expected outcome of 8. Further computations reveal that $f(9) = 40,329, f(10) = 181,450$, and $f(11) = 604,811$, so the assertion not only fails but fails spectacularly. A conjecture that worked perfectly for $n = 1, 2, 3, 4, 5, 6$, and 7 turned out to be quite incorrect.

The polynomial in question was generated by multiplying out and collecting terms in the expression

$$f(n) = n + [(n-1)(n-2)(n-3)(n-4)(n-5)(n-6)(n-7)]$$

Clearly, for $n = 1$, the term $(n-1)$ will be zero, thereby wiping out the entire product in square brackets; hence $f(1) = 1 + 0 = 1$. If $n = 2, n - 2 = 0$ and so $f(2) = 2 + 0 = 2$. Similarly, $f(3) = 3 + 0 = 3$, and so on up to $f(7) = 7 + 0 = 7$. But thereafter the term in brackets is no longer obliterated, so that, for instance, $f(8) = 8 + 7! = 5,048$.

This suggests a provocative extension. Suppose we introduced

$$g(n) = n + [(n-1)(n-2)(n-3) \ldots (n-1,000,000)]$$

and conjectured that $g(n) = n$ for all positive integers n.

Were we to multiply and collect the terms of $g(n)$, we would be left with a stupendous equation of the one-millionth degree. By reasoning identical to that above, we would find $g(1) = 1$, $g(2) = 2$, all the way up to $g(1,000,000) = 1,000,000$.

Having found a *million* consecutive confirmations of the conjecture, what sane person would doubt that $g(n)$ always yields n? For anyone—except a mathematician—1 million straight successes would constitute proof beyond all reasonable doubt. And yet the very next try—$g(1,000,001)$—turns out to be $1,000,001 + (1,000,000)!$, which is inconceivably larger than the expected 1,000,001.

This should reinforce the first maxim about mathematical proof: We must verify a result for *all* possible cases, not just for a few million.

Maxim #2: Simpler Is Better

Mathematicians admire proofs that are ingenious. But mathematicians especially admire proofs that are ingenious and economical—lean, spare arguments that cut directly to the heart of the matter and achieve their objectives with a striking immediacy. Such proofs are said to be elegant.

Mathematical elegance is not unlike that of other creative enterprises. It has much in common with the artistic elegance of a Monet canvas that depicts a French landscape with a few deft brushstrokes or a haiku poem that says more than its words. Elegance is ultimately an aesthetic, not a mathematical, property.

As with any ideal, elegance is not always achieved. Mathematicians strive for short, incisive proofs but often must settle for something cumbersome, tedious, and mind-numbing. For instance, a *single* proof from abstract algebra that classified what are called finite simple groups ran to more than 5,000 pages (when last anyone checked). Those seeking elegance are advised to look elsewhere.

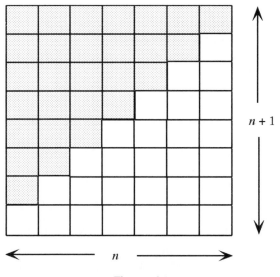

$n + 1$

n

Figure J.2

By contrast, an ultimate elegance is achieved by what mathematicians call a "proof without words," in which a brilliantly conceived diagram conveys a proof instantly, without need even for explanation. It is hard to get more elegant than that.

Consider, for example, the following:

THEOREM: If n is a positive integer, then $1 + 2 + 3 + \ldots + n = \dfrac{1}{2} n(n + 1)$.

This says that when we add the first n positive numbers, the sum is always half of the product of n and $n + 1$. We can easily check this for a few specific values of n; for instance, if $n = 6$,

$$1 + 2 + 3 + 4 + 5 + 6 = 21 = \frac{1}{2} (6 \times 7)$$

but our first maxim warns that only fools jump to conclusions based on a single case. Instead, we prove this proposition with the diagram in Figure J.2.

Here we have taken a "stairstep" configuration made up of one block plus two blocks plus three blocks, etc.; replicated it as the shaded section in Figure J.2; and fit them together to form an $n \times (n + 1)$ rectangular array. Because the rectangle is made of two identical stairsteps and the rectangle's area is the product of base and height—that is, $n(n + 1)$—then the stairstep's area must be half of the rectangle's. That is,

$$1 + 2 + 3 + \ldots + n = \frac{1}{2} n(n + 1)$$

as claimed. ■

The reader may observe (properly) that this "proof without words" was accompanied by a paragraph of explanation. But the verbal explanation was really unnecessary—the picture *was* worth a thousand words.*

Here is another proof with an undeniable elegance. Suppose we begin with 1 and add the consecutive odd integers:

$$1 + 3 + 5 + 7 + 9 + 11 + 13 + \ldots$$

A little experimentation suggests that, no matter how far we carry the sum, the result is always a perfect square. For example,

$$1 + 3 + 5 = 9 = 3^2$$
$$1 + 3 + 5 + 7 + 9 = 25 = 5^2$$
$$1 + 3 + 5 + 7 + 9 + 11 + 13 + 15 + 17 + 19 + 21 + 23 + 25 + 27 = 196 = 14^2$$

Is this always true? If so, how do we prove the general result?

The following argument, requiring a bit of algebra, rests upon the observation that an even number is a multiple of 2 and thus takes the form $2n$ for some integer n, whereas an odd number is one less than a multiple of 2 and looks like $2n - 1$ for some whole number n.

THEOREM: The sum of consecutive odd integers starting at 1 is a perfect square.

PROOF: Let S be the sum of consecutive odd integers from 1 to $2n - 1$. That is,

$$S = 1 + 3 + 5 + 7 + \ldots + (2n - 1)$$

It is clear that we will obtain S by taking the sum of *all* the integers from 1 to $2n$ and then subtracting the sum of the even ones. In other words,

$$S = [1 + 2 + 3 + 4 + 5 + \ldots + (2n - 1) + 2n] - [2 + 4 + 6 + 8 + \ldots + 2n]$$
$$= [1 + 2 + 3 + 4 + 5 + \ldots + (2n - 1) + 2n] - 2[1 + 2 + 3 + 4 + \ldots + n]$$

where we have factored a 2 out of the second bracketed expression.

The first set of square brackets contains the sum of all whole numbers from 1 to $2n$, and the second contains the sum of all whole numbers from 1 to n. The "proof without words" in Figure J.2 showed how to sum such strings of integers, so we apply the result twice:

$$S = \frac{1}{2} 2n(2n + 1) - 2\left[\frac{1}{2} n(n + 1) \right]$$

Upon simplifying we conclude

$$S = n(2n + 1) - n(n + 1) = 2n^2 + n - n^2 - n = n^2$$

---

* "Proofs without words" is a regular feature in the *College Mathematics Journal*.

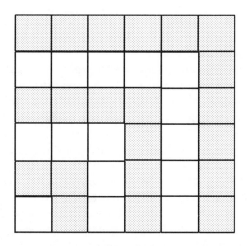

Figure J.3

Thus, regardless of the value of n, the consecutive odd integers sum to the perfect square n^2. The proof is complete. ■

That was, in a word, elegant. But if it is elegance we seek, Figure J.3 provides an even shorter alternative, another proof without words. Here are the odd integers—one block, three blocks, five blocks, and so on—arranged in a special way. We begin with a single block in the lower left-hand corner; three shaded blocks surround it to form a 2×2 square; five unshaded blocks surround these to form a 3×3 square; with the next seven shaded blocks we have a 4×4 square; and so on. The diagram makes clear that the sum of consecutive odd integers will always yield a (geometric) square. There is something very natural about this proof. It was known to the Greeks 2,000 years ago and can be mimicked by any modern youngster with building blocks.

It was Winston Churchill who remarked, "Short words are best and the old words when short are best of all."[7] This elegant argument can be paraphrased: Old proofs are best and the old proofs when short are best of all.

Maxim #3: Counterexamples Count

There is a stark fact of life in mathematics: To prove a general statement requires a general argument; to disprove it requires but a single specific instance in which the statement fails. The latter is called a ***counterexample,*** and a good counterexample is worth its weight in gold. Suppose, for example, we are given

CONJECTURE: If a and b are positive numbers, $\sqrt{a^2 + b^2} = a + b$

Over the years hundreds of thousands of students have invoked this very formula, as any math teacher will confirm. But it is fallacious, and to show this we need a counterexample. For instance, if $a = 3$ and $b = 4$, then $\sqrt{a^2 + b^2} = \sqrt{3^2 + 4^2} = \sqrt{25} = 5$, whereas $a + b = 3 + 4 = 7$. This lone counterexample is enough to send the conjecture to the mathematical landfill.

We stress that although it might require a 50-page argument to prove a theorem, a one-line counterexample will disprove it. In the great struggle between proof and disproof, it seems we don't have a level playing field. But a word of warning is in order: Finding counterexamples is not as easy as it might seem. The following story is a case in point.

More than two centuries ago Euler speculated that at least three perfect cubes must be added to get a sum that is itself a perfect cube, that at least four perfect fourth powers must be added to end up with a perfect fourth power, that at least five fifth powers must be summed to get a fifth power, and so on.

As an example, we add the cubes $3^3 + 4^3 + 5^3 = 27 + 64 + 125$ to get a sum of 216, which is just 6^3. Here *three* cubes combine to produce a cube, but Euler asserted—and proved—that *two* cubes were never adequate to sum to a perfect cube. The reader of Chapter F should recognize this as a special case (for $n = 3$) of Fermat's last theorem.

Moving up a degree, we can find four perfect fourth powers that sum to a fourth power, as in the considerably less trivial example

$$30^4 + 120^4 + 272^4 + 315^4 = 353^4$$

Euler conjectured that *three* fourth powers were never sufficient but he provided no proof. Generally he said that it takes at least n perfect nth powers to add up to another nth power.

So the matter stood in 1778, and so it stood nearly two centuries later. Those who believed Euler could not confirm his conjecture with a proof, but those who disbelieved him were unable to concoct a specific counterexample. The matter remained open.

Then in 1966 mathematicians Leon Lander and Thomas Parkin discovered that

$$27^5 + 84^5 + 110^5 + 133^5 = 61{,}917{,}364{,}224 = 144^5$$

Here a perfect fifth power arose as the sum of only *four* fifth powers. Euler was disproved. Then, two decades later a powerful computer flexed its electronic mind for a hundred hours before finding the even more remarkable counterexample:

$$95{,}800^4 + 217{,}519^4 + 414{,}560^4 = 422{,}481^4$$

This shows that three fourth powers, not Euler's four, could produce a fourth power.[8]

The amount of effort necessary to find these counterexamples—even when the effort came from computers—was staggering. This suggests a corollary to maxim #3: Sometimes disproving is harder than proving.

Maxim #4: You Can *Prove a Negative*

In barber shops or at lunch counters, one often hears the old adage that you can't prove a negative. Perhaps it comes up as follows:

A: "I read in a supermarket tabloid that a leprechaun won the lottery."

B: "There's no such thing as a leprechaun."

A: "What are you saying?"

B: "I'm saying leprechauns don't exist."

A: "Are you sure? Can you *prove* they don't exist?"

B: "Well . . . , no. But you can't prove a negative."

It's a great line. In one pithy utterance it declares that we cannot absolutely certify the nonexistence of elves or of Elvis.

Mathematicians know better. Some of the greatest, most profound mathematical arguments demonstrate that certain numbers, certain shapes, certain geometric constructions do not and cannot exist. And such nonexistence is established using the most incisive weapon of all: cold, hard logic.

At the core of the common notion that negatives are unprovable is a misconception. To prove the nonexistence of leprechauns seems to require us to search under every rock in Ireland and every iceberg in Antarctica. This, of course, is impossibly ambitious.

To establish nonexistence logically, mathematicians adopt a very different, yet perfectly sound, strategy: Assume the object *does* exist and track down the consequences. If we can show that the assumption of existence leads to a contradiction, then the laws of logic permit us to conclude that we erred in assuming existence in the first place. We can draw the incontrovertible conclusion of nonexistence, and the fact that we took an indirect route makes this conclusion no less valid.

In Chapter Q we consider the most famous nonexistence proof of all: why there exists no fraction equal to $\sqrt{2}$. However, for our present purposes the following simple example will suffice.

THEOREM: There exists no quadrilateral with sides of length 2, 3, 4, and 10.

A practical attack on the problem is to cut sticks of these lengths and try to arrange them into a four-sided figure. This is illuminating, yet in a logical sense it amounts to looking for a leprechaun under a rock. Even if we spend years unsuccessfully trying to form a quadrilateral from our sticks, that does not eliminate the possibility that someone, someday, might rearrange them into a four-sided figure.

Rather, we shall prove a negative indirectly. We begin—and here comes the great strategic leap—by assuming that there *is* a quadrilateral with sides 2, 3, 4, and 10 and then try to generate a contradiction.

Our supposed quadrilateral is depicted in Figure J.4. Draw the dotted diagonal, splitting it into two triangles, and let x be the diagonal's length. As shown in Chapter

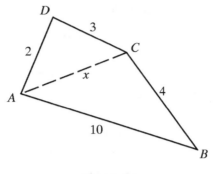

Figure J.4

G, Euclid proved that any side of a triangle is shorter than the sum of the other two. Thus, in $\triangle ABC$, we know $10 < 4 + x$. The same principle applied to $\triangle ADC$ yields $x < 2 + 3$. Combining these inequalities gives

$$10 \;<\; 4 + x \;<\; 4 + (2 + 3) = 9,$$

from which it follows that $10 < 9$. This is absurd. Our initial assumption about the existence of the specific quadrilateral led to this contradiction, and so we reject our assumption as invalid.

Here the sides of the quadrilateral appeared in the (clockwise) order 10, 2, 3, and 4. There are other ways to configure the sides—one is shown in Figure J.5—but similar reasoning still leads to a contradiction. Here, $10 < 2 + x < 2 + (3 + 4) = 9$, another impossibility.

There is no need to look further, no point in rearranging any more sticks. The quadrilateral in question cannot exist. We have conclusively proved a negative.

Proof by contradiction is a wonderful logical tactic. By assuming the opposite of what we intend to show, we seem to be putting the eventual goal in jeopardy. Yet, in the end, calamity is averted. G. H. Hardy (1877–1947) described proof by contradiction as "one of a mathematician's finest weapons. It is a far finer gambit than any chess gambit: a chess player may offer the sacrifice of a pawn or even a piece, but a mathematician offers *the game*."[9]

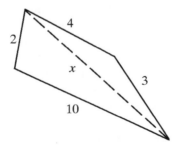

Figure J.5

Question: Are Humans Necessary?

During the 1970s and 1980s, an unsettling image entered the mathematical consciousness. It is the image of computers, with their lightning speed and virtual infallibility, taking over the job of proving theorems.

We have mentioned cases in which the computer provided counterexamples to *disprove* a statement. The discovery that $95,800^4 + 217,519^4 + 414,560^4 = 422,481^4$ dealt a death blow to Euler's conjecture, and it is difficult to imagine how long it might have taken human, as opposed to mechanical, intelligence to find this counterexample. Here we have the kind of problem ideally suited to a computer.

More troubling to the mathematical community are some recent instances in which computers were employed to *prove* a theorem. These tend to be done by shattering the theorem into a host of subcases, confirming each subcase, and then declaring the matter resolved. Unfortunately, the analysis often requires so many hundreds of cases and hundreds of millions of calculations that no one could possibly replicate all the steps; the proof, in short, can be checked only by another machine.

In 1976 the question of computer proofs burst dramatically upon the mathematical scene with the resolution of the four-color conjecture. This was the assertion that any map drawn upon a plane surface can be colored with four (or fewer) colors so that no two regions sharing a common boundary are assigned the same color. (In Figure J.6, for instance, we would not want to color regions A and B red, for then their common boundary line would be obliterated. We do allow two regions that intersect in a single point—such as regions A and C—to be colored the same; a point, of course, is not a boundary line.)

The four-color conjecture originated in 1852 and attracted much interest during the next century. A few matters were quickly resolved. It was proved that any planar map can certainly be colored with *five* colors. On the other hand, there are maps for which *three* colors are insufficient. One of these appears in Figure J.7. Here we must make regions A, B, and C of different colors because they share pairwise boundaries, but then it would be impossible to color region D unless a fourth color were introduced.

So five colors are plenty but three are too few. Obviously it all comes down to four. Are four enough to color *any* planar map?

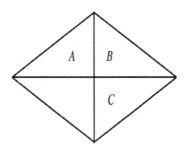

Figure J.6

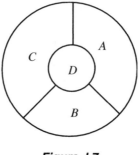

Figure J.7

Our previous discussion suggests that anyone wishing to resolve this issue has one of two options: Either come up with a specific counterexample—that is, a particular map—that cannot be colored with four colors or else devise a general proof that any map can be so colored. For mathematicians a counterexample proved elusive. Every map they created, no matter how intricate and convoluted, could be colored with just red, yellow, blue, and green. (The reader with a box of crayons may wish to scribble out an arbitrary map and give it a try.)

But, as we have noted repeatedly, proof requires more than just a few failed counterexamples. A general argument was aggressively sought but turned out to be every bit as difficult as any counterexample. The situation was at standstill.

Then Kenneth Appel and Wolfgang Haken of the University of Illinois shocked the mathematical world by announcing that the four-color conjecture was true. What was shocking was not the conclusion but the technique of proof: A computer had done the hard part.

Appel and Haken had approached the problem by dividing all planar maps into certain types and then analyzing each type separately. Unfortunately, there were hundreds upon hundreds of types to check, and each gave a good workout to a high-speed computer. In the end the computer announced that, yes, all possible types were colorable with just four colors. The theorem was proved.

Or was it? It is fair to say that an uneasiness spread across the mathematical world. Did this amount to a valid justification? The trouble was that it would take a genuine, flesh-and-blood human being maybe 100,000 years of 60-hour work weeks to check the computer's calculations. Even the healthiest, most optimistic person does not expect to be around that long, and anyway who would pay the overtime?

What if the programmers made a mistake? What if a power surge caused the machine to skip a critical step? What if the computer hardware had a subtle design flaw that would show up only in the rarest of cases? In short, do we trust a mechanical brain to give us the truth? As mathematician Ron Graham asked when mulling over this complicated issue, "The real question is this: if no human being can ever hope to check a proof, is it really a proof?"[10]

At this point the question has no definitive answer, although as computer proofs become more common mathematicians will probably feel more comfortable about their presence. It is fair to say, however, that most mathematicians would breathe a sigh of relief if the four-color theorem were established with a two-page proof—short, ingenious, and elegant—rather than with the brute-force machinations of a computer. Traditionalists long for the good old days of mathematics unplugged.

"Are humans necessary?" At the moment the answer is still, "Yes." After all, someone has to turn on the air conditioning. But we concede that this opinion may be biased because its proponents are, in fact, human.

This brings us to the end of our discussion of mathematical justification. Obviously, much more could be said, additional issues could be raised, and additional maxims could be advanced. But we conclude with perhaps most important observation of all: The standard of proof in mathematics—whether the proof be elegant or cumbersome, direct or indirect, computer or human generated—is a standard unmatched in any other field of human endeavor.

nighted Newton

On April 16, 1705, England's Queen Anne knighted Isaac Newton in a solemn Cambridge ceremony. With this act the queen bestowed Britain's highest honor upon one of its foremost subjects.

Newton shares with Gottfried Wilhelm Leibniz the distinction of creating the calculus, and in mathematics it is difficult to imagine a more illustrious distinction. This chapter sketches Newton's life, discusses his bitter dispute with Leibniz, and considers one of his mathematical legacies: the "method" that bears his name. (Our only regret is that, for the purposes of alphabetical ordering, he wasn't named Isaac Knewton.)

A striking fact about Sir Isaac is that he has become a kind of demigod for not one but two disciplines. Any mathematician, asked to name history's three or four most influential mathematicians, will include Newton among the elite. Any physicist, asked to identify history's three or four great physicists, is sure to include Newton as well.

It is true that Newton worked at a time before unscalable walls were erected between disciplines. In his day mathematics and physics were largely indistinguishable, with common methods, problems, and practitioners. It was a time when subjects such as optics, astronomy, and mechanics were treated as branches of mathematics. From today's vantage point, where mathematicians and physicists often find themselves so specialized as to be unable to communicate with one another, it is not easy to envision the situation three centuries ago when the boundaries were so blurred as to be nonexistent. In this light, one might minimize Newton's cross-disciplinary prominence.

Yet this would miss the point. To be held in such high esteem by two disciplines is extraordinarily rare. A parallel might be Shakespeare, as playwright and poet, or Michelangelo, as painter and sculptor, but their dual prominence may be no match for Newton's. His status is truly remarkable.

For Isaac, life had a precarious beginning. Born prematurely in Woolsthorpe, England, in 1642, he was given little chance of survival. In addition, his father had died some months before. But a greater trauma lay in store. In his third year Newton's widowed mother remarried and moved off to her new husband's home. Quite deliberately she left Isaac behind. She returned to him some years later, but many psychohistorians believe the damage had already been done. "All Newton's recent biographers," says a popular book on the subject, "have seen this separation from his mother, between the ages of three and ten, as crucial in helping to form the suspicious, neurotic, tortured personality of the adult Isaac Newton."[1]

Neurotic or not, the boy showed indisputable signs of genius. This ability, along with his obvious indifference to becoming a gentleman farmer, dictated that he should head off to university. He thus entered Trinity College, Cambridge, in the summer of 1661 and embarked upon an extraordinary scholarly career.

Making this intellectual journey possible was the fact that the professors at Cambridge were as indifferent to teaching as was Newton to the gentle art of farming. He was thus free to follow his interests, and these soon turned away from the heavy doses of Greek and Latin that characterized the official curriculum of the day and toward the exciting mathematical and scientific advances of the age. A solitary scholar, Newton devoured these subjects until, while still an undergraduate, he was doing original research. His work continued during the 1665–1667 hiatus when Cambridge was twice shut down because of outbreaks of the plague. Because of this Newton had to return to the family home in Woolsthorpe, although he hardly treated his homecoming as a leisurely vacation.

It was at Woolsthorpe that Newton had his famous encounter with the apple. According to legend, as he rested under a tree he was nearly hit by the falling fruit. He mused that if the earth tugged upon an apple, did it not also tug upon more distant celestial bodies? Newton recalled, "I began to think of gravity extending to ye orb of the Moon," which is about as concise an introduction to universal gravitation as one can ask.[2]

Modern scholars regard the falling apple as not so much a near miss as a near myth, but the story has its own charm. It prompted Lord Byron to write of Newton,

> this is the sole mortal who could grapple,
> Since Adam, with a fall, or with an apple.[3]

As the illustration shown here suggests, the apple has been stamped upon the public imagination as a symbol of Newton's extraordinary powers.

When the plague subsided, Newton returned to Trinity. In 1669, although still quite young and largely unknown, he assumed the prestigious Lucasian Chair of Mathematics at Cambridge. His great public breakthrough came in 1687 when, at the prodding of Edmund Halley, Newton finally agreed to publish something big—

Newton's apple on British postage

his *Principia Mathematica*. This work presented Newtonian mechanics in a precise, careful, and mathematical fashion. In it he introduced the laws of motion and the principle of universal gravitation and deduced, mathematically, everything from tidal flows to planetary orbits. *Principia Mathematica* is regarded by many as the greatest scientific book ever written.

With this triumph Newton entered the scientific limelight. Of course, the public had only the vaguest understanding of the details, but, rather like Einstein in the twentieth century, Newton became a living symbol of the new science. Voltaire

Isaac Newton
(Courtesy of Yerkes Observatory, University of Chicago)

called Newton "the greatest man that ever lived" and remarked that a genius of Newtonian proportions comes along only once in a thousand years.[4]

After his emergence from obscurity, Newton's life changed dramatically. In 1689 he served in Parliament as a representative from Cambridge. In 1696 he became warden of the Mint and moved to London for the remainder of his life. He was elected president of the Royal Society in 1703 and published his other great masterpiece, the *Opticks*, the following year. By the time of his death in 1727, Sir Isaac Newton was a revered scientist, a wealthy government servant, and an English hero worthy of burial alongside the elite in Westminster Abbey.

For mathematicians his most heroic discovery, dating to the mid-1660s, was the subject he called "fluxions" but that has come down to us bearing Leibniz's name of "calculus." For reasons that never seem entirely comprehensible from a modern perspective, Newton did not publish his discovery. In possession of perhaps the greatest mathematical innovation in history, he chose to keep quiet.

His strange and secretive nature did not serve him well. Time and again during his career, Newton would find others following the same intellectual paths he had traveled years before. Belatedly he would make a public claim of being the first discoverer, and this naturally would ruffle feathers across the scholarly world. It would have been so simple for him to have communicated his work at the time of discovery and thereby to have guaranteed not only his influence but also his reputation.

An explanation as to why he disdained publication seems always to come back to personality quirks: his distrust of others, his distaste of criticism, his "desire to decline being involved in such troublesome & insignificant Disputes."[5] Newton's views were concisely presented in his observation, "There is nothing wch I desire to avoyde in matters of Philosophy more then [sic] contention, nor any kind of contention more then one in print."[6]

What we have, therefore, is a scientist jealous of his own reputation yet reluctant to publicize his discoveries. Even with manuscripts prepared for private circulation, Newton sought to control distribution. "Pray let none of my mathematical papers be printed," he wrote a colleague in possession of an unpublished manuscript, "wthout my special licence."[7]

It does not take a genius of Newton's magnitude to predict that such behavior would have unfortunate repercussions. As time passed he became embroiled in priority controversies, nasty battles with other scientists about who did what when. He butted horns with countrymen Robert Hooke and John Flamsteed, but by far his most famous dispute was the controversy with Leibniz over the creation of the calculus.

With the hindsight of history, the basic facts of the case are these:

1. Newton had discovered his method of fluxions by the mid-1660s. He described it in a 1669 manuscript known as *De analysi* and an expanded 1671 treatise called *De methodis fluxionum*. These circulated among a select group of British mathematicians but were not published and hence not widely known. Those who read them instantly recognized Newton's power, and one described him as "very young . . . but of an extraordinary genius and proficiency."[8]

2. During the mid-1670s, a full decade later, Leibniz made virtually the same discoveries. While on a diplomatic mission to London in 1676, Leibniz saw a manuscript copy of Newton's *De analysi*.

3. At about the same time, Leibniz received two letters from Isaac—what have come to be called the *epistola prior* and the *epistola posterior*—revealing some of Newton's thoughts about infinite series and, much less explicitly, about fluxions.

4. In 1684 Leibniz published the first paper on differential calculus, as we noted at the beginning of Chapter D. Nowhere in it did he mention that he had seen manuscripts or exchanged letters with Newton eight years earlier. In fact, nowhere did he mention Newton at all.

This is not to suggest that Leibniz plagiarized Newton (although this is precisely the suggestion many English mathematicians advanced). The manuscript trail establishes that Leibniz, his contacts with Newton notwithstanding, had discovered the principles of the calculus independently and rightfully shares the glory of discovery. And because of Newton's chronic secretiveness, Leibniz's 1684 paper was unquestionably the source from which the scholarly world learned of this wonderful new subject.

Obviously both parties had made mistakes. Newton could have published his research at any time during the two decades between its discovery and Leibniz's paper, and the question of priority would have been moot. With his silence, Newton asked for trouble. For his part, Leibniz could have acknowledged his contact with Newtonian documents and more generously shared the credit that, he knew, deserved to be shared. By *his* silence, Leibniz let the world believe he was the lone discoverer. Being less than forthright came back to haunt him as the dispute grew more heated.

Soon after Leibniz's 1684 publication, Newton started grumbling about priority, and these grumbles developed into an thinly veiled rage. In Newton's opinion, only *first* discoverers deserved recognition (even if those discoverers took enormous pains to conceal their work from public view).[9] In 1699, the texts of Newton's two 1676 letters to Leibniz were published, and the British believed they had found a "smoking gun"—or, in the jargon of the day, a "smoking blunderbuss"—to convict the latter of scholarly theft.

After that, the situation spiraled downward into the muck. Accusations became too numerous to follow without a scorecard, and loyal subordinates joined the two main characters in lobbing verbal grenades back and forth across the Channel. To us it seems incredibly unbecoming, but ours is the perspective of distance, uncontaminated by the passions of the day and not subject to the divisions of nationality that separated the British and their continental rivals.

As an indication of the tone of the exchanges, we offer one grenade from each side. A British follower of Newton penned this 1708 statement, imprudently published in the *Philosophical Transactions* of the Royal Society:

> All of these [conclusions] follow from the now highly celebrated Arithmetic of Fluxions which Mr. Newton, beyond all doubt, First Invented, as anyone who reads his Letters . . . can easily determine; the same Arithmetic under a different name and using a different notation was later published in the *Acta eruditorum*, however, by Mr. Leibniz.[10]

Although a lawyer could argue that plagiarism was not explicitly charged here, the reference to Newton's ideas being "later published . . . , however, by Mr. Leibniz" using "different notation" made the intent fairly clear. Leibniz certainly thought so.

He complained loudly to the Royal Society for having endorsed such offensive remarks.

It was a complaint he lived to regret, for in response the Society organized a committee to investigate the priority dispute. Their report, published in 1713 under the title *Commercium epistolicum*, supported Newton on all counts. Leibniz, it suggested, had no inkling of the calculus until the middle of 1677, long after he had received Newton's letters and seen Newton's manuscripts. The conclusion of *Commercium epistolicum* was inescapable: Leibniz had stolen the ideas of the master. This harsh verdict, however, loses some of its impact when one recognizes that Newton was then president of the Royal Society and wrote much of the *Commercium* himself.

Charge and countercharge continued. Soon an anonymous broadside appeared on the Continent with a stridently pro-Leibniz stance. In it one finds this passage:

> when Newton took to himself the honour due to another of the analytical discovery of differential calculus first discovered by Leibniz . . . he was too much influenced by flatterers ignorant of the earlier course of events and by a desire for renown; having undeservedly obtained a partial share in this . . . he longed to have deserved the whole—a sign of a mind neither fair nor honest.[11]

Here we read that it was *Newton* who unfairly stole Leibniz's thunder and not vice versa. Such a ridiculous charge, of course, is the price Newton paid for his refusal to publish. It may come as no surprise that the author of this anonymous attack was later revealed to be Gottfried Wilhelm Leibniz.

In retrospect, the mutual denunciations of two of the greatest mathematicians of all time make for a sad chapter in European intellectual history. That individuals of such genius descended to petty and outrageous mudslinging does not bode well for those of us with more modest intellects. The whole matter is a colossal embarrassment for Newton, for Leibniz, for mathematics, and for scholarship generally.

This unseemly dispute tends to tarnish Newton's image. To a lesser extent, so, too, does his preoccupation, over a period of decades, with alchemy and theology.

Alchemy, of course, was the medieval pursuit in which scientist/magicians tried to change common chemicals into gold. Newton, who read voluminously on the subject, spent huge amounts of time at the furnaces he had built, diligently heating chemicals and watching for that golden glimmer. Although he seemed even more secretive about his alchemy than about his fluxions, his alchemical notes eventually totalled nearly a *million* words.

His theological writings were similarly vast. Newton was a master at scrutinizing the Bible, discerning prophecies, connecting seemingly unconnected passages. His notes include a floor plan for the Temple in Jerusalem that he concocted from scriptural passages, and he published works such as *Observations upon the Prophecies of Daniel, and the Apocalypse of St. John* (in two parts). This was clearly a subject to which he gave his foremost attention.

Unfortunately, whereas both mathematics and physics have been forever enriched by Newton's work, his legacy as a theologian is nonexistent, and alchemists are now regarded as akin to snake-oil salesmen. It is tempting to wonder what additional scientific fruit might have fallen at Newton's feet had he devoted less time to these matters.

We now turn to one matter that certainly *was* worthy of his genius: the so-called "Newton's method" for approximating solutions of equations. The form in which we describe it is not precisely the one discovered by Newton in the 1660s. His technique, modified by Joseph Raphson in 1690 and Thomas Simpson in 1740, has come down to us in a somewhat different guise. But even with modification, the essential idea is his.

The issue at hand is one of the most fundamental in all of mathematics: to solve an equation. Many mathematical journeys eventually lead to this very end, yet the procedures of algebra are limited in their ability to provide exact solutions. For instance, the quadratic formula shows that the solutions of $7x^2 - 24x - 19 = 0$ are

$$\frac{12 + \sqrt{277}}{7} \quad \text{and} \quad \frac{12 - \sqrt{277}}{7}$$

but no algebraic technique will yield exact answers for

$$x^7 - 3x^5 + 2x^2 - 11 = 0$$

What if we need a solution for just such an equation? What does a mathematician do when confronted with an unsoluable problem?

The strategy is to aim a little lower. If the exact answer is not available, try to find an approximate one. After all, a solution accurate to ten decimal places would be sharp enough for any practical need. Moreover, if the approximation technique were fairly simple, if it carried with it its own theoretical underpinnings, and if it could be used repeatedly to get ever more accurate estimates, then the procedure would be nearly as good as the exact solution itself. Fortunately, these properties characterize Newton's method.

Before proceeding, we observe that the two equations above were written with zero on the right-hand side. This was no accident, for we stipulate that equations be put in that format before applying the method. Of course, this is easily achieved by moving all terms to the left. That is, rather than dealing with $x^3 + 3x = 7x^5 - x^2 + 2$, we transfer terms across the equals sign to get the equivalent

$$-7x^5 + x^3 + x^2 + 3x - 2 = 0$$

and thereby reach the desired form $f(x) = 0$. We shall insist upon this in all that follows.

At this point, we need a bit of geometric insight. Consider the graph of $y = f(x)$ as shown in Figure K.1. Solving the equation $f(x) = 0$ amounts to finding the value of x at which the graph intersects the x-axis. Such a point is called the **x-intercept** of the function and is designated by c in the figure. If we can determine c (at least approximately), we shall have solved the equation $f(x) = 0$ (at least approximately).

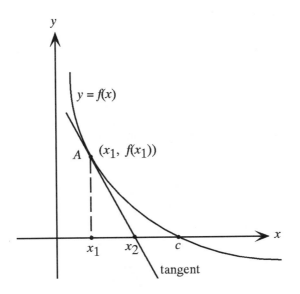

Figure K.1

Newton's method requires that we begin with a guess about the solution. In Figure K.1 we labeled our first guess as x_1. In essence we are saying that $x_1 \approx c$, the actual solution. The drawing shows that the estimate is not terribly good, for x_1 is considerably smaller than c, but not to worry. The genius of Newton's method is that it provides a scheme for improving the estimate with each use.

Starting at x_1 on the horizontal axis, we look directly above to the corresponding point A on the curve $y = f(x)$. This point has coordinates $(x_1, f(x_1))$. As indicated, we draw the tangent line to the curve at A. It is here that differential calculus enters the picture, for we recall from Chapter D that the *slope* of this tangent is the derivative of the function evaluated at $x = x_1$. Symbolically, the slope of the tangent is $f'(x_1)$.

Now imagine descending along the curve $y = f(x)$ from left to right. Ideally we would continue along the sweeping descent until arriving at the exact value c that is the solution to the equation. But because this exact solution is unknown, we instead leave the curve at the point A and move downward along the tangent line. The point x_2 where the tangent line intersects the x-axis, although not exactly the point c, is at least a closer approximation to c than was our initial guess of x_1.

The previous paragraph contains the geometric essence of Newton's method. But how do we determine the new estimate x_2 algebraically? The answer is to consider the slope of the tangent from two different viewpoints and equate results. As noted, the slope of the tangent line is given by the derivative $f'(x_1)$. On the other hand, the slope of *any* line is found by the expression:

$$\text{slope} = \frac{\text{rise}}{\text{run}} = \frac{y_2 - y_1}{x_2 - x_1}$$

As the diagram indicates, the tangent line passes through $(x_1, f(x_1))$ and $(x_2, 0)$. Its slope is therefore

$$\frac{0 - f(x_1)}{x_2 - x_1} = -\frac{f(x_1)}{x_2 - x_1}$$

Equating these two expressions for slope, we solve for x_2:

$$f'(x_1) = \text{slope of tangent line} = -\frac{f(x_1)}{x_2 - x_1}$$

and so

$$x_2 - x_1 = -\frac{f(x_1)}{f'(x_1)}$$

which implies that

$$x_2 = x_1 - \frac{f(x_1)}{f'(x_1)}$$

We have thus found an expression for x_2, our improved estimate of c, based on (1) the value of x_1, our previous guess; (2) the value of the function f at x_1; and (3) the value of the derivative f' at x_1. Of course, we still do not know the exact value of c, but with this formula we have refined the approximation.

What if x_2 is not accurate enough for our tastes? We simply apply the whole argument again, this time starting with x_2. This yields yet a better estimate

$$x_3 = x_2 - \frac{f(x_2)}{f'(x_2)}$$

as indicated in Figure K.2. From the diagram it appears that our approximate solution x_3 and the true solution c differ by very little. But, of course, we can apply the procedure again. In general, if x_n is the approximate solution at step n, then the next approximation is

$$x_n - \frac{f(x_n)}{f'(x_n)}$$

and this formula embodies what is known as Newton's method.

An example or two are in order. Suppose we wish to approximate $\sqrt{2}$. As we shall see in Chapter Q, no ten-place, or ten-million-place, decimal will give it exactly. Yet we often need an estimate of $\sqrt{2}$ accurate to a goodly number of places.

These days, one simply uses a calculator and lets the machine do the computation. In a sense, however, this begs the question, for how does the calculator find $\sqrt{2}$? Put another way, how could a mere mortal come up with an answer?

The best approach is to apply Newton's method. We first note that $\sqrt{2}$ is the solution of the quadratic equation $x^2 = 2$ or, equivalently, $x^2 - 2 = 0$; here we have adopted the format $f(x) = 0$ with $f(x) = x^2 - 2$. In Chapter D we showed that the

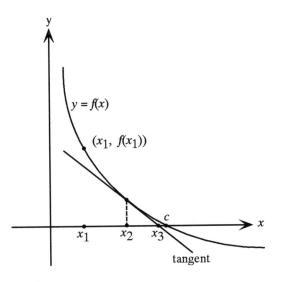

Figure K.2

derivative of x^2 is $2x$, and any constant has derivative (i.e., slope) of zero. Thus $f'(x) = 2x - 0 = 2x$.

Newton's method then says that if x_1 is our first approximation to the solution of $x^2 - 2 = 0$, then our second is

$$x_2 = x_1 - \frac{f(x_1)}{f'(x_1)} = x_1 - \frac{x_1^2 - 2}{2x_1}$$

If we find a common denominator and simplify, this becomes

$$x_2 = \frac{2x_1^2 - (x_1^2 - 2)}{2x_1} = \frac{x_1^2 + 2}{2x_1}$$

An analogous argument applied to the approximation x_n yields as next approximation

$$\frac{x_n^2 + 2}{2x_n}$$

It now only remains to make a first guess for $\sqrt{2}$. A reasonable choice would be $x_1 = 1$. We then repeatedly apply Newton's method:

$$x_2 = \frac{x_1^2 + 2}{2x_1} = \frac{1 + 2}{2} = \frac{3}{2}$$

$$x_3 = \frac{x_2^2 + 2}{2x_2} = \frac{(9/4) + 2}{3} = \frac{17/4}{3} = \frac{17}{12}$$

$$x_4 = \frac{x_3^2 + 2}{2x_3} = \frac{(289/144) + 2}{17/6} = \frac{577/144}{17/6} = \frac{577}{144} \times \frac{6}{17} = \frac{577}{408}$$

$$x_5 = \frac{x_4^2 + 2}{2x_4} = \frac{(332,929/166,464) + 2}{577/204} = \frac{665,857}{470,832}$$

Converting these to decimals yields the string of approximations

$$x_1 = 1.000000000\ldots$$

$$x_2 = 1.500000000\ldots$$

$$x_3 = 1.416666666\ldots$$

$$x_4 = 1.414215686\ldots$$

$$x_5 = 1.414213562\ldots$$

As a matter of fact, to nine places $\sqrt{2} = 1.414213562\ldots$, so four repetitions of Newton's method has yielded nine-place accuracy. Moreover, such a reiteration scheme, in which the output of one step is the input of the next, is perfectly adapted to what programmers call a "loop." It makes Newton's method quick and efficient on a computer.

Our other example is Newton's own. In the 1669 treatise (unpublished, of course) in which he first described the method, he addressed the cubic equation $x^3 - 2x - 5 = 0$. To approximate a solution, we set $f(x) = x^3 - 2x - 5$ so that $f'(x) = 3x^2 - 2$ by the differentiation rules of Chapter D. Then Newton's method tells us that if we have x_n as the current approximation to the solution, the next one will be

$$x_n - \frac{f(x_n)}{f'(x_n)} = x_n - \frac{x_n^3 - 2x_n - 5}{3x_n^2 - 2} = \frac{2x_n^3 + 5}{3x_n^2 - 2}$$

Here a reasonable first guess is $x_1 = 2$, as $f(2) = 2^3 - 2(2) - 5 = -1$, which is fairly close to 0. Applying the recursion three times gives:

$$x_1 = 2$$

$$x_2 = \frac{2(2^3) + 5}{3(2^2) - 2} = \frac{21}{10} = 2.1$$

$$x_3 = \frac{2(2.1)^3 + 5}{3(2.1)^2 - 2} = \frac{23.522}{11.23} = 2.094568121$$

$$x_4 = \frac{2(2.094568121)^3 + 5}{3(2.094568121)^2 - 2} = \frac{23.37864393}{11.16164684} = 2.094551482$$

So our approximate solution is $x = 2.094551482$. Upon substitution into the original cubic, we find $x^3 - 2x - 5 = (2.094551482)^3 - 2(2.094551482) - 5 = 0.000000001$, which is about as close to 0 as anyone could want. Newton's method, repeated just three times, zeroed in on the answer simply and efficiently. He himself seemed justifiably pleased with the technique and wrote: "I do not know whether this method of

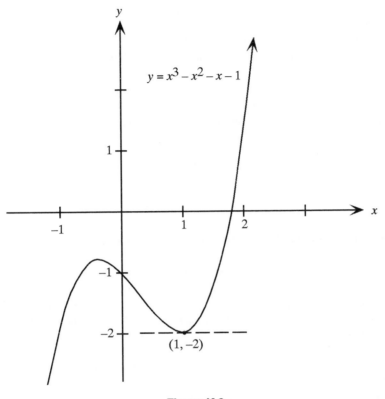

$$y = x^3 - x^2 - x - 1$$

$(1, -2)$

Figure K.3

resolving equations is widely known or not, but certainly in comparison with others it is both simple and suited to practice . . . and [it] is easily recalled to mind when needed."[12]

In all fairness, we should issue a word of warning: In spite of the previous numerical examples, there are times when the use of Newton's method requires a bit more care. Consider, for instance, the cubic equation $x^3 = x^2 + x + 1$. Proceeding as above, we take all terms to the left side and write it as $f(x) = x^3 - x^2 - x - 1 = 0$. The derivative rules from Chapter D tell us that $f'(x) = 3x^2 - 2x - 1$.

Suppose we now choose $x_1 = 1$ as a first estimate and substitute into the key formula to get:

$$x_2 = x_1 - \frac{f(x_1)}{f'(x_1)} = 1 - \frac{f(1)}{f'(1)} = 1 - \frac{-2}{0}$$

But division by 0 is *never* allowed in this or any other mathematical procedure. The expression $-2/0$ has no meaning. Newton's method has failed.

If we return to the original theory, it is easy to see what went wrong. In Figure K.3 we have sketched the graph of $y = f(x) = x^3 - x^2 - x - 1$ and the first guess of

$x_1 = 1$. Because $f(1) = -2$, we go down to the point $(1, -2)$, draw the tangent line, and let the next guess x_2 be the point where the tangent line intersects the x-axis. But here the tangent is *horizontal* and thus parallel to the x-axis. Because the tangent and x-axis never meet, the point of intersection x_2 required by Newton's method simply fails to exist.

Fortunately the snag is easily fixed. One of the wonderful features of Newton's method is that it contains its own corrective medicine. We need only make a *different* initial guess, such as $x_1 = 2$, and let the method churn out the approximations:

$$x_2 = 1.857142857$$
$$x_3 = 1.839544512$$
$$x_4 = 1.839286812$$
$$x_5 = 1.839286755$$

Sure enough, $x = 1.839286755$ satisfies the original cubic equation with great precision.

Today there is an important and very useful branch of mathematics called numerical analysis that looks at the finer points of approximation procedures. The subject has become very subtle and very deep, but its trademark is Newton's method. This is one of the great theorems of mathematics and one of the most far-reaching applications of differential calculus.

We close with a last word about Isaac Newton and his remarkable mathematical career. As noted, he was not without personality flaws, and some scholars even regard his secretive, neurotic behavior as a sign of madness. But, to paraphrase the words of Shakespeare: Though this be madness, yet there is (Newton's) method in't.

 ost Leibniz

As noted in the previous chapter, Isaac Newton is ranked among the greatest mathematicians of all time. His achievements were numerous, but standing above all others is his creation of the calculus.

This is an honor he shares with his contemporary Gottfried Wilhelm Leibniz. In fact it was Leibniz who gave the subject its distinctive notation and even its name. Yet the same scholars who put Newton atop the list *because of* his creation of calculus regularly overlook Leibniz *in spite of* his creation of calculus. Somehow, Leibniz seems to have gotten lost. This is not only unfair but unfortunate, for in many ways his story is as remarkable as Newton's.

Gottfried Wilhelm Leibniz was born in 1646 in Leipzig. Even as a child, his reading demonstrated a broad range of interests, and he seemed to possess the ability to learn just about anything with amazing speed. Leibniz must have been a most impressive scholar when he entered university at the age of 15. Three years later he had earned bachelor's and master's degrees, and soon thereafter he received a doctorate in law from the University of Altdorf. The world seemed to lie at the feet of this brilliant and engaging young man.

Meanwhile, in Cambridge, Newton was working night and day on his marvelous fluxions. But Leibniz, although accomplished in so many disciplines, knew little of mathematics at this point in his life. "When I arrived in Paris in the year 1672," he recalled decades later, "I was self-taught as regards geometry, and indeed had little knowledge of the subject, for which I had not the patience to read through the long series of proofs."[1] Even Euclid was largely a mystery to him, and when he chanced to look into Descartes's *Géométrie,* he found it much too difficult.[2] No one would have guessed that within a few years his discoveries would place him among the giants of mathematics.

It was the law that occupied Leibniz for the better part of the next decade. He was employed as an advisor to the elector of Mainz and in this capacity embarked on a diplomatic mission to Paris in March of 1672. This proved to be the great experience of his life. The young diplomat was dazzled by the artistic, literary, and scientific energy he found there. He fell in love with Paris and all it represented during this, the reign of the Sun King.

Among the intellectuals who inhabited the French capital, none exerted a more profound influence on Leibniz than the Dutch scientist and mathematician Christiaan Huygens (1629–1695). Huygens, who served as a sort of mentor during this crucial period, wanted to gauge his young friend's mathematical acumen and thus challenged Leibniz to determine the sum of the infinite series

$$1 + \frac{1}{3} + \frac{1}{6} + \frac{1}{10} + \frac{1}{15} + \frac{1}{21} + \frac{1}{28} + \frac{1}{36} + \dots$$

(Here the denominator of the *n*th fraction is the sum of the first *n* whole numbers.)

Leibniz, having to rely on raw intelligence rather than past training, experimented a bit before rewriting the series as

$$1 + \frac{1}{3} + \frac{1}{6} + \frac{1}{10} + \frac{1}{15} + \frac{1}{21} + \frac{1}{28} + \dots = 2\left[\frac{1}{2} + \frac{1}{6} + \frac{1}{12} + \frac{1}{20} + \frac{1}{30} + \frac{1}{42} + \frac{1}{56} + \dots\right]$$

Then, expressing each fraction within the square brackets as a difference of two others, he transformed the right-hand side into

$$2\left[\left(1 - \frac{1}{2}\right) + \left(\frac{1}{2} - \frac{1}{3}\right) + \left(\frac{1}{3} - \frac{1}{4}\right) + \left(\frac{1}{4} - \frac{1}{5}\right) + \left(\frac{1}{5} - \frac{1}{6}\right) + \left(\frac{1}{6} - \frac{1}{7}\right) + \dots\right] = 2[1] = 2$$

because within the brackets all terms after the initial 1 cancel. In this fashion, he correctly concluded that

$$1 + \frac{1}{3} + \frac{1}{6} + \frac{1}{10} + \frac{1}{15} + \frac{1}{21} + \frac{1}{28} + \frac{1}{36} + \dots = 2$$

The mathematical novice had passed Huygens's test. Historian Joseph Hoffman, commenting on the critical role played by this problem in Leibniz's career, observed that "another example only slightly more difficult (and hence for Leibniz insoluble) would no doubt have quenched his enthusiasm for . . . mathematics."[3] Instead, success ignited him.

Leibniz did more than solve one problem. Fascinated by infinite series, he considered many other examples and later observed that an investigation of such sums was central to his discovery of the calculus.[4] As was to become his mathematical trademark, Leibniz sought an underlying principle that unified a wide class of similar problems. In large measure, his genius lay in the ability to discern general rules that linked specific and seemingly unrelated examples. It takes a penetrating intellect to accomplish such a synthesis, and this Leibniz certainly possessed.

A second characteristic of his work was an appreciation of good notation. He advocated an "alphabet of thought," a collection of symbols and rules that, if followed, would ensure correct reasoning not only in mathematics but in everyday life. Although this grandiose plan never came close to fruition, it is regarded as a precursor of modern symbolic logic. And if Leibniz did not succeed in symbolizing all of human thought, he introduced the notation of calculus that is used to this day.

In Paris, his intellectual odyssey accelerated. As was his custom, he read voraciously and, although his diplomatic efforts must have suffered as a consequence, he moved rapidly toward the mathematical frontier. By the spring of 1673, he was making discoveries of his own. "I was now ready to get along without help," Leibniz remembered, "for I read [mathematics] almost as one reads tales of romance."[5]

Some of these discoveries now are regarded as minor curiosities. For instance, he solved a challenge problem to find three numbers whose sum is a perfect square and the sum of whose squares is a square of a perfect square (such arcane problems were popular in his day). Leibniz found the numbers 64, 152, and 409, whose sum $64 + 152 + 409 = 625 = 25^2$, a perfect square, and the sum of whose squares is

$$64^2 + 152^2 + 409^2 = 194{,}481 = (441)^2 = (21^2)^2$$

the square of a square. *How* he found these is not important at this moment, but we stress: It was not by guessing.[6] Leibniz also discovered the bizarre formula

$$\sqrt{1 + \sqrt{-3}} + \sqrt{1 - \sqrt{-3}} = \sqrt{6}$$

which not only perplexed some of the world's great mathematicians (including, in a sense, him) but helped popularize the imaginary numbers that are the subject of Chapter Z.[7]

All of this was merely an overture for the great performance of Leibniz's mathematical career. Working in his Paris rooms, he pushed his research ever deeper so that, by the autumn of 1675, he was in possession of the "new method," the subject we now know as calculus. It was an exhilarating time for him and a momentous one for mathematics. When modern visitors walk the streets of Paris, they are apt to think of the art, the music, the literature that was created in that great city, and people such as Victor Hugo or Toulouse-Lautrec seem to live again. But few realize that these same avenues saw the birth of calculus more than three centuries ago. If Paris has spawned great art, it has also spawned great mathematics. That so few are aware of this is another indication of how badly Leibniz has gotten lost.

Gottfried Wilhelm Leibniz
(Courtesy of Lafayette College Library)

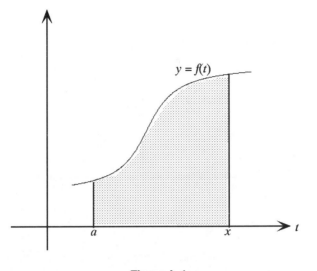

Figure L.1

His diplomatic mission lasted from 1672 until the autumn of 1676 when he had to return to his native Germany. It was there that he published the first account of differential calculus in 1684. Two years later a second paper introduced the other branch of the subject, integral calculus, which occupies us for the remainder of this chapter.

As we have seen, differential calculus deals with the slopes of curves. On the other hand, integral calculus addresses the areas beneath them. By considering *area*, integral calculus thus takes aim at a problem whose roots go back thousands of years.

Our discussion begins with a generic function whose graph lies above the horizontal axis. The goal of integral calculus is to determine the shaded area under the curve $y = f(t)$ between any two points along this axis, say, from $t = a$ on the left end to $t = x$ on the right as shown in Figure L.1. (In what follows we use t rather than x as the independent variable, a notational convention that will prove useful.)

We have previously found areas enclosed by such figures as circles (Chapter C) or trapezoids (Chapter H). But there we needed a different formula for each different figure. Integral calculus, by contrast, adopts a more general viewpoint and seeks a unified method to find areas bounded by arbitrary functions. This is a far more ambitious goal.

A reasonable starting point is to recall the advice that when faced with something unknown, try first to relate it to something known. Thus we approach the irregular shaded area by means of the areas of simpler, better known figures—in this case, ordinary rectangles.

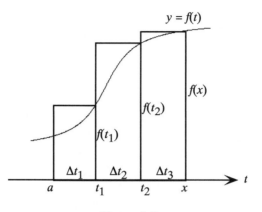

Figure L.2

That is, divide the horizontal interval between a and x into three smaller segments, called *subintervals,* at the points t_1 and t_2, as shown in Figure L.2. We denote the *lengths* of the three subintervals by:

$$\Delta t_1 = t_1 - a, \Delta t_2 = t_2 - t_1, \text{ and } \Delta t_3 = x - t_2$$

Next, construct a rectangle upon each subinterval. Of course, not just any rectangle will do, for it must bear some relation to the curve $y = f(t)$. So choose the *height* of the rectangle upon the interval from a to t_1 to be the value of the function at t_1. In symbols, the height of the left-hand rectangle is $f(t_1)$. Consequently the *area* of this rectangle is (height) × (base) = $f(t_1) \Delta t_1$. Similarly the middle rectangle has height $f(t_2)$ and area $f(t_2) \Delta t_2$, and the right-hand rectangle has height $f(x)$ and area $f(x) \Delta t_3$.

We have thereby approximated the area under the original curve by the *sum* of the areas of the three rectangles. That is,

area under curve ≈ sum of rectangular areas = $f(t_1) \Delta t_1 + f(t_2) \Delta t_2 + f(x) \Delta t_3$

Obviously we have here a very crude approximation of the exact area shaded in Figure L.1. How can it be improved?

Pretty clearly the trick is to take more, and thinner, rectangles. In Figure L.3 we have split the interval from a to x into not three but six pieces, of widths Δt_1, $\Delta t_2, \ldots, \Delta t_6$, and six skinny rectangles are erected upon them. As a consequence, we have

area under curve ≈ sum of rectangular areas = $f(t_1) \Delta t_1 + f(t_2) \Delta t_2 + \ldots + f(x) \Delta t_6$

which is an improvement because the narrower rectangles more nearly approximate the exact area under the curve.

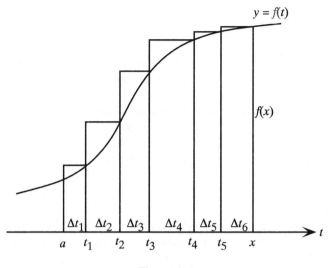

Figure L.3

Why stop with six? Adopting a general perspective, divide the interval from a to x into n pieces, of widths $\Delta t_1, \Delta t_2, \ldots, \Delta t_n$, put rectangles upon each, and get the approximation

area under curve $\approx$ sum of rectangular areas $= f(t_1) \Delta t_1 + f(t_2) \Delta t_2 + \ldots + f(x) \Delta t_n$

The larger the n, the thinner the rectangles and the better job they do of estimating the area in question. But even a thousand narrow rectangular strips with not give the exact area under a curve. To get the area precisely, we must turn to the idea of limits.

Recall that limits appeared in Chapter D, where they played the key role in the definition of the derivative. This time limits are the pivotal idea behind the integral. Rather than stopping with a thousand—or a million—rectangles, we let their number grow without bound even as their widths shrink away toward zero. In so doing, we shall have determined the area under the curve. That is,

area under curve $= \lim [f(t_1) \Delta t_1 + f(t_2) \Delta t_2 + \ldots f(x) \Delta t_n]$

where we take the limit as the lengths of all the subintervals tend to zero. In passing to the limit we can replace $\approx$ by $=$ and drop the qualifiers about "approximating" area; when the limit has been taken, the resulting area is *exact*.

As was his custom, Leibniz introduced a new symbol. He denoted the area beneath a curve with "∫", an elongated letter "S" for "sum," to suggest the summation of rectangular areas. Interestingly, we know the date on which he chose this notation: October 29, 1675.[8] Ever afterward the area under $y = f(t)$ between $t = a$ and $t = x$ has been denoted by

$$\int_{a}^{x} f(t)dt$$

This is the ***integral,*** defined above as the limit of the sum of rectangular areas, and the process of finding integrals is called ***integration.*** It is unquestionably one of the fundamental concepts of higher mathematics.

At this point an example is in order. Suppose we wish to find the area beneath the straight line $y = f(t) = 2t$ from $t = 0$ to $t = 1$, as shaded in Figure L.4. Here the region is a simple triangle, so we can find its area without recourse to integral calculus. The triangle is one unit wide and two units high, so its area is simply

$$\frac{1}{2}bh = \frac{1}{2}(2 \times 1) = 1$$

As an alternative we determine the same area—and presumably get the same answer—with an integral. Figure L.5 shows what happens when the interval from 0 to 1 is divided into five equal subintervals and the associated rectangles are introduced.

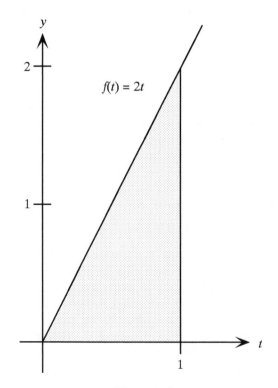

Figure L.4

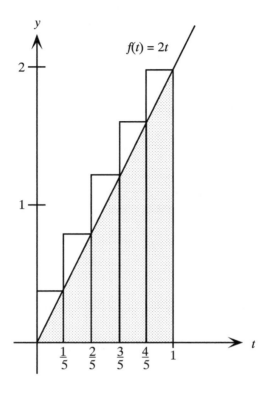

Figure L.5

Certainly the sum of the five rectangular areas exceeds the triangular area we seek, but at least it provides a first approximation. Each rectangle has base of length 1/5 and their heights are, respectively,

$$f\left(\frac{1}{5}\right)=\frac{2}{5}, \quad f\left(\frac{2}{5}\right)=\frac{4}{5}, \quad f\left(\frac{3}{5}\right)=\frac{6}{5}, \quad f\left(\frac{4}{5}\right)=\frac{8}{5}, \text{ and } f(1)=2$$

Thus

$$\text{sum of rectangular areas} = \left(\frac{1}{5}\times\frac{2}{5}\right)+\left(\frac{1}{5}\times\frac{4}{5}\right)+\left(\frac{1}{5}\times\frac{6}{5}\right)+\left(\frac{1}{5}\times\frac{8}{5}\right)+\left(\frac{1}{5}\times 2\right)$$

$$=\frac{2}{25}+\frac{4}{25}+\frac{6}{25}+\frac{8}{25}+\frac{10}{25}$$

$$=\frac{2}{25}(1+2+3+4+5)$$

$$=\frac{2}{25}(15)=\frac{6}{5}=1.20$$

which, as expected, overestimates of the exact triangular area of 1.

Notice that in the next to last line of this derivation we encountered the sum of the first five positive integers. In fact, if we instead divide the interval from 0 to 1 into n equal pieces, an exactly parallel argument shows that

$$\text{sum of rectangular areas} = \left(\frac{1}{n} \times \frac{2}{n}\right) + \left(\frac{1}{n} \times \frac{4}{n}\right) + \left(\frac{1}{n} \times \frac{6}{n}\right) + \ldots + \left(\frac{1}{n} \times 2\right)$$

$$= \frac{2}{n^2}(1 + 2 + 3 + \ldots + n)$$

where we now must sum the first n integers. Fortunately, our "proof without words" in Chapter J showed that the sum in the parentheses is

$$\frac{n(n + 1)}{2}$$

We then substitute to get

$$\text{sum of rectangular areas} = \frac{2}{n^2}(1 + 2 + 3 + \ldots + n) = \frac{2}{n^2} \times \frac{n(n + 1)}{2}$$

$$= \frac{n^2 + n}{n^2} = \frac{n^2}{n^2} + \frac{n}{n^2} = 1 + \frac{1}{n}$$

In words, this says that the sum of the areas of the n rectangles exceeds 1 by an amount equal to $1/n$.

Of course, n rectangles will never give the precise area in question. So we take the limit as n tends toward infinity to get the exact area:

$$\int_0^1 2t\,dt = \lim_{n \to \infty}(\text{sum of rectangular areas}) = \lim_{n \to \infty}\left(1 + \frac{1}{n}\right) = 1$$

because $1/n$ approaches zero as the n in its denominator grows ever larger.

Here is the answer we found using the geometric formula above. Integral calculus took a very roundabout path to reach the same result. But what is significant is that our geometric formula was specific to triangles, whereas the ideas of integration are applicable to figures vastly more complicated. With integral calculus we can determine areas under parabolas, ellipses, and countless other curves well beyond the range of elementary geometry. It is the generality of the method that makes it so powerful.

Unfortunately as our functions become more sophisticated, the process of summing rectangular areas and taking limits becomes extremely complicated. If our goal is to determine areas in an automatic and relatively painless way, a shortcut is essential. In Paris in the mid-1670s, Gottfried Wilhelm Leibniz found it.

The shortcut has become known as the ***fundamental theorem of calculus,*** a result whose very name suggests its overwhelming significance. The theorem is fun-

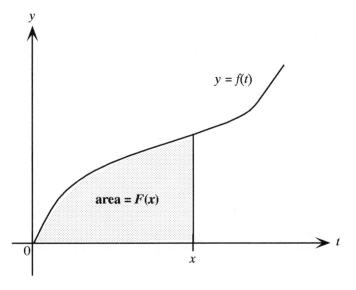

Figure L.6

damental not only because it turns the evaluation of areas into an easy problem but also because it links the apparently unrelated concepts of derivative and integral. The theorem is thus the great connector between the two branches of calculus.

Return to the general curve $y = f(t)$. Consider the shaded area beneath it between $t = 0$ and $t = x$, as shown in Figure L.6 (our choice of left-hand endpoint at 0 reflects a common seventeenth-century practice and makes what follows a bit simpler). We let $F(x)$ represent this area. That is, in Leibniz's notation,

$$F(x) = \int_0^x f(t)dt$$

Note that F is actually a function of x, for as x moves to the right, $F(x)$, the shaded area under the curve between 0 and x, will grow as well. The function F is simply an "area accumulator" function whose value depends on how far to the right x is placed.

The goal is to find some kind of formula for F. Such a formula would allow us to determine the area

$$\int_0^x f(t)dt$$

merely by substituting x into F. Integration would be automatic *if* we knew the identity of F.

How do we find it? Strangely enough, the trick is not to attack F directly but to attack its *derivative*. That is, we shall determine $F'(x)$ and from this infer the for-

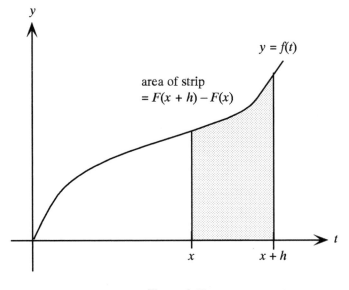

Figure L.7

mula for F itself. It may seem hopelessly indirect, but the outcome is more than worth the indirectness.

At this point the reader may wish to glance back to Chapter D, where the derivative was introduced. According to that definition, the derivative of F is:

$$F'(x) = \lim_{h \to 0} \frac{F(x+h) - F(x)}{h}$$

So, we take a small value of h. By the definition of F, we know that $F(x + h)$ is the area under the curve $y = f(t)$ between $t = 0$ and $t = x + h$ just as $F(x)$ is the area under the curve between $t = 0$ and $t = x$. Consequently, $F(x + h) - F(x)$ (which appears in the numerator of the derivative expression above) is the difference of these areas; in short, $F(x + h) - F(x)$ is the area of the shaded strip in Figure L.7.

Generally it is impossible to determine the exact area of this strip, because its top is bounded by a portion of the irregular curve $y = f(t)$. We are thus forced to *approximate* the strip's area.

To do this, draw the straight line connecting $(x, f(x))$ and $(x + h, f(x + h))$ shown in Figure L.8. The result is a trapezoid. Its bases are of length $f(x)$ and $f(x + h)$ and its height (the distance between the parallel bases) is h. Therefore, by the trapezoidal area formula developed in Chapter H, we find

$$\text{area(trapezoid)} = \frac{1}{2} h\, (b_1 + b_2) = \frac{1}{2} h\, [f(x) + f(x + h)]$$

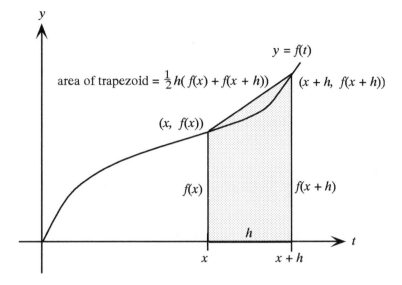

Figure L.8

Now we use this *trapezoidal* area to approximate the area of the irregular strip in Figure L.7. That is, if F is the area accumulator function, then

$$F(x+h) - F(x) = \text{area of irregular strip} \approx \text{area of trapezoid} = \frac{1}{2} h \, [f(x) + f(x+h)]$$

It follows that

$$\frac{F(x+h) - F(x)}{h} \approx \frac{\frac{1}{2} h \, [f(x) + f(x+h)]}{h} = \frac{f(x) + f(x+h)}{2} \qquad (*)$$

Finally, to determine the derivative $F'(x)$, we take the limit of this expression as $h \to 0$. In so doing, the difference between the true area of the strip and that of the trapezoidal approximation vanishes. Moreover, provided that the original function f is reasonably well behaved, we find that as $h \to 0$, then $f(x+h) \to f(x+0) = f(x)$. Combining all of this at last brings us to the fundamental theorem of calculus:

$$F'(x) = \lim_{h \to 0} \frac{F(x+h) - F(x)}{h} \qquad \text{by definition of the derivative}$$

$$= \lim_{h \to 0} \frac{f(x) + f(x+h)}{2} \qquad \text{by } (*) \text{ above}$$

$$= \frac{f(x) + f(x)}{2} \qquad \text{because } f(x+h) \to f(x)$$

$$= \frac{2f(x)}{2} = f(x)$$

It is time to stop and get our bearings. What exactly has been accomplished by this long argument?

First, recall the original objective: to find a simple expression for

$$F(x) = \int_0^x f(t)dt$$

What we have discovered is not the identity of F but that of its derivative. And, remarkably, $F'(x)$ turned out to be just $f(x)$, the function that bounded the area we sought.

To put it differently, we needed the area under $y = f(t)$. We started with f, integrated it to get F, and then differentiated F (that is, differentiated the integral) only to end up at f again. The fundamental theorem of calculus says that the derivative of the integral of the function f is f. Just as addition undoes subtraction and division undoes multiplication, so, too, does differentiation undo integration. The two great ideas of calculus are thus joined. Differentiation and integration are two sides of the same coin.

We finish up with two examples to indicate that all this work was worth the effort. First, return to the triangular area in Figure L.4, where we had to evaluate

$$\int_0^1 2t\,dt$$

In this case $f(t) = 2t$ and we let

$$F(x) = \int_0^x 2t\,dt$$

The fundamental theorem of calculus says that $F'(x) = f(x) = 2x$. In other words, F is the function whose derivative is $2x$. But in Chapter D we explicitly showed that x^2 is the function whose derivative is $2x$. We conclude that $F(x) = x^2$.

From here, the area of Figure L.4 is easily found. We know that

$$\int_0^x 2t\,dt = F(x) = x^2$$

and, replacing x by 1, we get

$$\text{area of triangle} = \int_0^1 2t\,dt = F(1) = 1^2 = 1$$

This is the answer we obtained twice before. Everything seems to be consistent.

As a second example look back to Chapter B and our discussion of the Monte Carlo method. There a probabilistic argument was used to estimate the area of a lake bounded by the curve $y = 8x - x^2$ and shown again in Figure L.9.

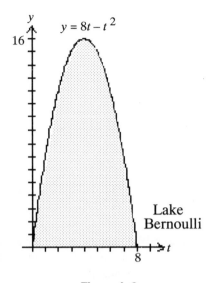

Figure L.9

With the ideas just developed we can determine the *exact* area of the lake. Note that this is the area beneath a parabola. Most people, even those who remember the area formula for a triangle or trapezoid, have no idea about the area of a parabolic segment. This is a job for integral calculus.

We denote the lake's area by

$$\int_0^8 (8t - t^2)dt$$

(where we use the variable t rather than x for the function forming the lakeshore). Introduce the area accumulator

$$F(x) = \int_0^x (8t - t^2)dt$$

In this problem, $f(t) = 8t - t^2$, so the fundamental theorem tells us that $F'(x) = f(x) = 8x - x^2$.

By recalling the derivative rules from Chapter D and thinking backwards—mathematicians call this *antidifferentiation*—we conclude that

$$F(x) = 4x^2 - \frac{1}{3}x^3$$

This follows because the derivative of

$$4x^2 - \frac{1}{3}x^3$$

is

$$4(2x) - \frac{1}{3}(3x^2) = 8x - x^2 = f(x)$$

Then by the fundamental theorem

$$\int_0^x (8t - t^2)\, dt = F(x) = 4x^2 - \frac{1}{3}x^3$$

and so the lake's area is found by letting $x = 8$:

$$\int_0^8 (8t - t^2)\, dt = F(8) = 4(8^2) - \frac{1}{3}(8^3) = 256 - \frac{512}{3} = 85.3333\ldots$$

Recall that the Monte Carlo method approximated the area of the lake as 84.301. This is reasonably close to the exact answer of 85.333 just obtained, suggesting that the law of large numbers and the fundamental theorem of calculus are working in concert.

We conclude this chapter with two words of warning. First, as the reader may have sensed, a complete theory of integration is *vastly* more complicated than we have here let on. Our development has been naive and nonrigorous, with carefully chosen examples and many logical gaps. In this sense it mirrors the early, unsophisticated thinking about integral calculus. As later mathematicians grappled with these ideas, they encountered theoretical obstacles of great subtlety, obstacles that were finally resolved only in the latter years of the nineteenth century.

Our other word of warning is more germane: Gottfried Wilhelm Leibniz deserves his share of the spotlight. By a trick of history he lived in the days of Isaac Newton, and if the bright star of Newton's genius has eclipsed Leibniz in the popular memory, it is arguable that Newton's star would eclipse that of anyone.

Nonetheless, the mathematical community owes Leibniz his due. Like Newton, he discovered the great ideas of differentiation and integration and recognized the fundamental theorem of calculus as the bridge between them; unlike Newton, he shared these ideas with a receptive world. In so doing, Leibniz inspired others, most notably the Bernoullis, and through their individual research and mutual correspondence, they crafted the subject as we know it today. *Our* calculus, in a very real sense, is the calculus of Leibniz.

When all is said and done, the critical fact is that, at such an important juncture in the history of mathematics, not one but two individuals of the highest caliber were simultaneously active: Isaac Newton and his peer Gottfried Wilhelm Leibniz.

Mathematical Personality

This chapter is awash in stereotypes. Heedless of the potential injustice of mis-characterizing an entire group of people or the very real possibility of getting sued, we insert tongue firmly in cheek and discuss commonly held views about the mathematical personality.

The average men or women in the street, when they think about mathematicians at all, tend to regard them as intelligent, abstract, hyperlogical, socially backward, preoccupied, withdrawn, myopic, or—to use the comprehensive adjective—nerdy. Is this a valid assessment? Do mathematicians really exhibit these character traits, or are they the victims of a widespread misconception?

Some years ago George Pólya, a Stanford professor highly respected as a mathematician and much beloved as a teacher, addressed this very issue. Drawing upon the experiences of a lifetime, Pólya identified two common characteristics: (1) Mathematicians are absentminded and (2) mathematicians are eccentric.[1] These provide a good starting point.

The charge of absentmindedness seems to be right on target. The folklore is rife with stories of mathematicians who are forever missing appointments, misplacing important papers, losing their eyeglasses. For example, there is the oft-repeated tale of Witold Hurewicz, a renowned mathematician, who drove his car to New York City, parked it and went about his business, and then took the train home. The next day, finding an empty parking space, Hurewicz phoned the police to report a stolen vehicle.[2]

Pólya relates the story of a young professor newly arrived at Göttingen University in the early years of the twentieth century. The novice wished to pay respects at the home of the esteemed mathematician David Hilbert. Dressed in his finest outfit, he knocked at Hilbert's door and was invited in for the perfunctory introduction. The young man removed his hat, took a seat, and began chattering. He quickly outlasted his welcome. Hilbert's attention wandered into some arcane mathematical problem. After some minutes of this, Hilbert decided he had had enough. He rose, picked up the young visitor's hat, politely said goodbye, and left. One can only guess the reaction of the visitor, sitting alone in the professor's living room.[3]

Tales of absentminded mathematicians certainly are not limited to the twentieth century. It was Archimedes who, having made a remarkable discovery while bathing, leapt naked from the bath and ran through town in the maximum state of excitement and, unfortunately, in the minimum state of clothing. We are told that Isaac Newton worked so diligently in his rooms he forgot to eat the meals that had been brought to him. On those occasions when he *did* walk to the dining hall, Newton "would go very carelessly, with Shooes down at Heels, Stockins unty'd, surplice on, & his Head scarcely comb'd."[4]

Then there is one of the nineteenth century's great absentminded mathematicians, Peter Gustav Lejeune Dirichlet of Germany. Dirichlet, who was Gauss's successor on the mathematics faculty at Göttingen, is often described as not just absentminded but "notoriously" absentminded. It is said that Dirichlet was so preoccupied that he forgot to tell his in-laws of the birth of their first grandchild. The grandfather, when he finally learned the news after an exasperating delay, observed that Dirichlet at least could have written "$2 + 1 = 3$."[5] Upon his death, Dirichlet's brain was removed for later study, surely an instance of carrying absentmindedness to an extreme.

These stories, and many others, suggest that absentmindedness is a chronic affliction of mathematicians. However, not everyone is convinced of this, so in the interest of fairness we briefly mention the opposing viewpoint of John F. Bowers of the University of Leeds. In a provocative article about the eccentricities of mathematicians, Bowers bucked the prevailing wisdom by declaring unequivocally, "the idea that mathematicians are absent-minded is absolutely wrong. There is a conclusive proof that shows that they are not, but unfortunately it cannot be given here because it seems to have been mislaid."[6]

That serious mathematicians suffer this malady is hardly surprising. After all, they grapple daily with the most abstract concepts, the most unforgiving logic, the most intractable challenges. The typical student finds it exhausting to devote an hour to a single problem, but how many can imagine devoting months or even years to such a task? The necessary concentration is awesome, and absentmindedness seems to be a natural consequence. It was the absentminded Newton who said he made a great discovery only by "thinking on it continually."[7]

When people spend years thinking continually about things such as the distribution of prime numbers or the trisection of angles, it is little wonder they neglect to comb their hair. The material world begins to seem so trivial, so arbitrary, so ephemeral when contrasted with the timeless beauty of mathematics. It is not surprising that mathematicians forget to put out the cat; indeed, they often forget they *own* a

Peter Gustav Lejeune Dirichlet
(Courtesy of Muhlenberg College Library)

cat. Their bodies may be resting comfortably in an armchair, but their minds are wandering through very different realms.

As noted above, Pólya also asserted that mathematicians were eccentric. This may be an open-and-shut case, for anyone who spends a lifetime thinking about those primes or trisections is automatically exhibiting a certain degree of eccentricity. Of course, on the surface most mathematicians act as normal as their banker or lawyer. But to the trained observer, certain clues give them away.

For one, there is attire. It seems clear that mathematicians select apparel with an eye toward comfort rather than style. Perhaps the utter absurdity of some conventions of fashion—men's neckties, for example—is particularly grating to the unwaveringly logical mathematician. They rarely will be found in silk dresses or gray flannel suits, preferring instead cotton shirts bearing inscriptions such as

$$\int_{0}^{\infty} e^{-x^2} dx = \frac{\sqrt{\pi}}{2}$$

For many, the footwear of choice is sandals with black socks, and for others, dressing up means putting on a *new* pair of sneakers.

Along these lines, we should mention the cartoon icon of a mathematician in white lab coat standing before a symbol-laden blackboard. Mathematicians do, in fact, spend an inordinate amount of time staring at symbol-laden blackboards. But *never* do they wear white lab coats. Such garments are as likely to be found on sumo wrestlers as on mathematicians. Cartoonists, take note.

There is little doubt that male mathematicians are disproportionately bearded. Full facial hair is the unofficial uniform of the profession, perhaps because shaving is illogical (if men were meant to have smooth faces, why would little hairs grow out of their chins?). The prevailing wisdom suggests that roughly 50 percent of male mathematicians are hirsute. The only place one is likely to encounter more beards is at a Santa Claus convention or during a curtain call for *Fiddler on the Roof.*

Then there are the eyeglasses. These are nearly universal. Of course, there are times when mathematicians absentmindedly misplace their spectacles, but by and large they can be seen gazing intently through their corrective lenses, although the object of their gaze might be an invisible equation or an unseen polygon.

Mathematicians are also known for a distinctive brand of humor, a kind often described as "dry," although "parched" may be more accurate. This in turn can be subdivided into two categories, which here are called "low" mathematical humor and "high" mathematical humor.

Low humor involves the intentional confusion of mathematical terminology. Over dozens of centuries mathematicians have generated a vast lexicon of technical terms. Some of these, such as *homotopy* or *diffeomorphism,* remain the exclusive possession of specialists. Others, such as *matrix* or *parameter,* have filtered into the general language, where they are misused on a daily basis. In yet other cases, everyday words have been borrowed and introduced into the mathematician's vocabulary. Thus there there are precise mathematical meanings for terms such as *field* or *group* or *pencil.*

All of this allows mathematicians to interchange the technical and common meanings of words to their heart's delight. They will call a gathering of colleagues a "finite group" and chuckle knowingly. They will describe a set of twins as being not identical but "isomorphic." When a situation is improving, mathematicians will say it has a "positive derivative."

Mathematicians also exploit sound-alikes. Everyone has heard jokes in which "hypotenuse" is replaced by the name of a large aquatic mammal. The number π probably holds the record for bad puns related to a baked confection (see the cartoon, Chapter C). And it was only with the greatest restraint that, in discussing the *Elements* in Chapter G, we avoided the overused but wonderful subtitle, "Here's looking at Eu-Clid."

Fortunately, there is a higher form of mathematical humor that goes beyond mere puns. This commonly involves distortions of logic. The humor arises, after a moment's thought, from some sort of logical inconsistency. Mathematicians, whose scholarship is logic-driven, find it especially funny when the wheels come off.

We begin with an example from Pólya. Looking back from late in his career, he reminisced about his lifelong affection for the discipline of philosophy and wrote: "Who is a philosopher? The answer is: A philosopher is one who knows everything and nothing else."[8] This quip has the sort of logical convolution mathematicians find amusing.

Along similar lines is a comment made by the physicist Wolfgang Pauli. Pauli, who displayed brilliance and arrogance in roughly equal proportions, once disparaged a new colleague with the wonderfully twisted comment, "He's so young and already so unknown."[9] Or consider Stephen Bock's description of a sheltered man and his dreams: "Reading was something Jay knew about only from books, yet he was quite anxious to experience it for himself."[10]

The use—or misuse—of logic also figures in the story of mathematician Henry Mann who, it is told, drove himself and a group of colleagues to scientific meetings in Cincinnati. Unfamiliar with Cincinnati streets, Mann became ever more lost. His colleagues, although uneasy, remained silent until they realized he had turned the wrong way onto a one-way street. But Mann dismissed their warnings. This street could not possibly be one-way, he pointed out, since *their* car was heading in one direction and any number of vehicles were coming toward them in the other.[11]

These are examples of logic turned on its head. The following story finds humor in the illogic of English pronunciation. The Polish mathematician Mark Kac immigrated to the United States and tried to master the sometimes inexplicable English language. Particularly vexing were those words that, although spelled with identical endings, were pronounced differently. For instance, the "ow" ending might be pronounced with a long O, as in *grow* or *know,* yet the same ending is said very differently when it appears in *cow* or *how.* Of course, the word *bow,* with two different pronunciations, incorporates the worst of both worlds.

In any case, Professor Kac, in grappling with this phenomenon, came to realize that the word *snowplow* was doubly bizarre, because the same "ow" is pronounced in two different ways *within the same word.* Mindful of this, he took special care to recall its illogical pronunciation. Unfortunately, he interchanged the variants, so that instead of rhyming *snowplow* with *grow-cow,* he rhymed it with *cow-grow.*[12]

Finally, there is this story with its own surprise twist. During an informal moment at a mathematical conference, a young admirer asked the noted mathematician R. H. Bing for his autograph. With this in hand, she asked Paul Halmos, another celebrated mathematician, to sign on the same sheet. Thus she held in her hand what was the mathematical equivalent of a page jointly autographed by Gilbert and Sullivan, or Ruth and Gehrig, or Siskel and Ebert.

When she showed this prize to a colleague, he immediately said, "I'll give you $25 for that page." But another, wittier mathematician jumped in with the clincher: "Yes, but *I'll* give you $50 if you let me sign my name below theirs."

These examples indicate the sort of humor popular among mathematicians. A moment's thought is required, and the typical response is not necessarily to laugh at the humor but to *appreciate* it. Neither ribald nor slapstick, mathematical humor tends to be cerebral. One suspects that very few mathematicians are numbered among the Three Stooges Fan Club.

If attire and humor, eccentricity and absentmindedness set mathematicians apart, their shared identity may be viewed as something of a defense mechanism. They truly find strength in numbers.

For instance, there is the widespread impression that mathematicians are merely accountants who spend their days adding up columns of numbers. Mathematician/poet JoAnne Growney, in confronting this perception, was moved to verse:

Misunderstanding

Ah, you are a mathematician,
    they say with admiration
    or scorn.

Then, they say,
    I could use you
    to balance my checkbook.

I think about checkbooks.
    Once in a while
    I balance mine,
    just like sometimes
    I dust high shelves.[13]

Are mathematicians misunderstood? Certainly. Are they scorned? Without a doubt. The two comments most often heard when someone is introduced to a mathematician are, "I hate mathematics," or "I fear mathematics," although these may be combined into the incomparable, "I hate *and* fear mathematics."

Why should mathematicians constantly be bombarded with such remarks? Why do so many people regard the subject as the academic equivalent of eye surgery without anesthetic? Were they, as children, bitten by a mathematician? Upon inquiry, one discovers two common sources of mathophobia: Either the respondent once had a terrible math teacher or the respondent perceives a terminal inadequacy in his or her mathematical abilities.

The former excuse, that of the bad instructor, is quite widespread and quite remarkable. People who forget things such as their wedding anniversary or the name of the president can nonetheless remember with absolute clarity an offending algebra teacher from decades ago. Whether Mr. Jones or Ms. Smith really was as awful as claimed or whether the bad memories have some deeper, darker origin is always a matter of speculation.

But if the math teacher from hell is one excuse shared by millions, even more common is the explanation, "I never could do math and I never will." This is a confession every teacher of mathematics has heard hundreds of times. It suggests that mathematical success is genetic. Just as some people are born with blue eyes, so others are born with the ability to do math. If you were not so born, you are destined to be a mathematical basket case, and nothing can change the prognosis.

It is not easy to disabuse people of this attitude. Many who encounter difficulty with mathematics quickly conclude that the fault lies in their stars and not in themselves. All too few draw the opposite inference, that a little more study might help.

So mathematicians hunker down in the face of this onslaught. Colleagues in other disciplines rarely encounter such attitudes. It is difficult to imagine the following exchange in a history class:

Professor: "George, who was president of the U.S. during the Civil War?"

George: "Um . . . um . . . um . . . I'm sorry, prof, I never could do history."

Unfortunately, some not only chant a mathophobic mantra but also *cherish* it. This is true even of otherwise highly educated people. If a mathematician bragged about never having read a lick of poetry, she or he would be branded as an ignorant lout. Yet the poet who admits to being mathematically illiterate often wears this illiteracy as a badge of pride. Something about this seems unfair.

With a lack of mathematical understanding comes the inability to recognize the true significance of mathematical ideas. Imagine the following scene:

We are at a cocktail party of learned men and women engaged in highbrow chitchat. Over by the piano, a biologist is describing to a rapt audience the feeding habits of the Komodo dragon, while around the sofa a heated discussion is raging about the bouquet of California wines. These topics are understandable not just to specialists but also to the general audience, even those who are not herpetologists or chefs.

There comes a lull in the conversation. A mathematician in the corner takes a sip of ginger ale, fumbles with a plastic pencil holder, and remarks that

$$\int_0^\infty e^{-x^2} dx = \frac{\sqrt{\pi}}{2}$$

Conversation ceases. Glasses stop clinking. There is deadly silence. People check their watches or reach for their coats. Many show signs of terror. The party is over.

In point of fact, the formula above

$$\int_0^\infty e^{-x^2} dx = \frac{\sqrt{\pi}}{2}$$

is not only true but also essential for our understanding of the normal probability distribution. The normal distribution, in turn, lies at the heart of statistical inference. Medical research, polling data, and scores of other important questions depend squarely upon the validity of this formula. As such, it is far more central to modern life than Komodo dragons or table wines. Yet few non-mathematicians have the slightest appreciation of the power contained in this string of symbols. Only other mathematicians fully "get it." As a group, they must deal as best they can with this lack of public understanding. It's a hard life.

And so, if you come upon a collection of bespectacled, bemused individuals, all of whom are talking seriously, some of whom are wearing socks with sandals, and none of whom are dressed in lab coats; and if they appear to be a finite group lounging around a trigonometric table making bad puns; and, further, if none of them think the Three Stooges are the least bit funny—then you can confidently place your bet: You are in the presence of mathematicians.

Please treat them with kindness.

Natural Logarithm

$$\ln(e^x) = x$$

This chapter tells the story of a special number—denoted by "e"—and its eternal partner, the natural logarithm. At first glance, neither seems either special or natural. Intuition, in fact, suggests that they are relatively insignificant. Our goal is to explain why, in this case, intuition is at fault.

We begin with e. Of course, "e" is the fifth letter of the English alphabet, but the mathematician's e is a real number with decimal expansion 2.718281828459045 . . . Whereas everyone knows that "e," the most frequently used letter in the English language, is indispensable, it may come as a surprise to non-mathematicians that e is likewise indispensable. Why should this peculiar number, just a bit smaller than 2-3/4 be any more important than, say, 2.12379 . . . or 3.55419 . . . or any other garden-variety decimal?

Before answering this question, we must explain how e is defined and calculated—in short, where it comes from. There are two different but logically equivalent sources, one involving limits and the other infinite series. We examine the limit definition first.

Consider the expression

$$\left(1 + \frac{1}{k}\right)^k$$

where k is a positive integer. If $k = 2$, we have

$$\left(1 + \frac{1}{2}\right)^2 = (1.5)^2 = 2.25$$

if $k = 5$, we get

$$\left(1 + \frac{1}{5}\right)^5 = (1.2)^5 = 2.48832$$

if $k = 10$,

$$\left(1 + \frac{1}{10}\right)^{10} = (1.1)^{10} = 2.59374\ldots$$

and so on. Mathematicians, always ready to push things to the limit, let k grow without bound and define

$$e = \lim_{k \to \infty}\left(1 + \frac{1}{k}\right)^k$$

In words, e is the limit of the kth power of the expression $1 + 1/k$ as the number k grows ever larger. With the assistance of a calculator we generate the first few places in the decimal expansion of e:

k	$1 + \dfrac{1}{k}$	$\left(1 + \dfrac{1}{k}\right)^k$
10	1.1	2.59374246 ...
100	1.01	2.70481383 ...
1,000	1.001	2.71692393 ...
1,000,000	1.000001	2.71828047 ...
1,000,000,000	1.000000001	2.71828183 ...
↓		↓
∞		e

Evidently, $e \approx 2.71828183$.

With a bit of work it is possible to prove the more general result:

FORMULA A: $\displaystyle \lim_{k \to \infty}\left(1 + \frac{x}{k}\right)^k = e^x$

Here the number x inside the parentheses becomes the exponent on e when we take the limit as $k \to \infty$. Note that if we let $x = 1$ in formula A, we return to the previous result

$$\lim_{k \to \infty}\left(1 + \frac{1}{k}\right)^k = e^1 = e$$

The other way to generate e is to sum the infinite series

$$e = 1 + \frac{1}{1!} + \frac{1}{2!} + \frac{1}{3!} + \frac{1}{4!} + \frac{1}{5!} + \frac{1}{6!} + \ldots$$

$$= 1 + 1 + \frac{1}{2} + \frac{1}{6} + \frac{1}{24} + \frac{1}{120} + \frac{1}{720} + \ldots$$

where the denominators involve factorials as introduced in Chapter B. The more terms we add in this series, the closer we approach the numerical value of e.

Of course, these two formulas for e look very different. However, it can be established that

$$\lim_{k \to \infty} \left(1 + \frac{1}{k}\right)^k = 1 + \frac{1}{1!} + \frac{1}{2!} + \frac{1}{3!} + \frac{1}{4!} + \frac{1}{5!} + \frac{1}{6!} + \ldots$$

In this sense it is instructive to evaluate

$$1 + \frac{1}{1!} + \frac{1}{2!} + \frac{1}{3!} + \frac{1}{4!} + \frac{1}{5!} + \frac{1}{6!} + \frac{1}{7!} + \frac{1}{8!} + \frac{1}{9!} + \frac{1}{10!} + \frac{1}{11!}$$

The sum turns out to be 2.71828183, precisely the approximation of e generated by the limit definition above.

Then, using the series approach, we find any power of e—in other words, e^x for any x—by means of

FORMULA B: $1 + \dfrac{x}{1!} + \dfrac{x^2}{2!} + \dfrac{x^3}{3!} + \dfrac{x^4}{4!} + \dfrac{x^5}{5!} + \dfrac{x^6}{6!} + \ldots = e^x$

To estimate e^2, for instance, we substitute $x = 2$ into formula B and add the first dozen or so terms of the series. This, in essence, is what a scientific calculator does when we push the number 2 followed by the e^x key and read the output: $e^2 = 7.389056099 \ldots$

In the history of mathematics the person most closely associated with e is Leonhard Euler, whom we met in Chapter E and elsewhere throughout this book. It was Euler who chose the symbol for this constant and who grasped its overwhelming importance. In Figure N.1, reproduced from his *Introductio in Analysin Infinitorum* of 1748, we see Euler introducing what we called formula B—except he wrote e^z rather than e^x—and providing the decimal expansion of e to an astonishing (in the days before computers) 23 places.[1]

We have described two ways to define and calculate this particular number. But why bother? Why is it important, and why is it *natural*? As we shall see, its uses are almost endless.

qui termini, si in fractiones decimales convertantur atque actu addantur, praebebunt hunc valorem pro *a*

$$2,71828\,18284\,59045\,23536\,028,$$

cuius ultima adhuc nota veritati est consentanea.

Quodsi iam ex hac basi logarithmi construantur, ii vocari solent logarithmi *naturales* seu *hyperbolici*, quoniam quadratura hyperbolae per istiusmodi logarithmos exprimi potest. Ponamus autem brevitatis gratia pro numero hoc 2,71828 18284 59 etc. constanter litteram

$$e,$$

quae ergo denotabit basin logarithmorum naturalium seu hyperbolicorum [1]), cui respondet valor litterae $k = 1$; sive haec littera *e* quoque exprimet summam huius seriei

$$1 + \frac{1}{1} + \frac{1}{1 \cdot 2} + \frac{1}{1 \cdot 2 \cdot 3} + \frac{1}{1 \cdot 2 \cdot 3 \cdot 4} + \text{etc. in infinitum.}$$

123. Logarithmi ergo hyperbolici hanc habebunt proprietatem, ut numeri $1 + \omega$ logarithmus sit $= \omega$ denotante ω quantitatem infinite parvam, atque cum ex hac proprietate valor $k = 1$ innotescat, omnium numerorum logarithmi hyperbolici exhiberi poterunt. Erit ergo posita *e* pro numero supra invento perpetuo

$$e^z = 1 + \frac{z}{1} + \frac{z^2}{1 \cdot 2} + \frac{z^3}{1 \cdot 2 \cdot 3} + \frac{z^4}{1 \cdot 2 \cdot 3 \cdot 4} + \text{etc.}$$

Figure N.1

Euler introduces *e*

(Courtesy of Lehigh University Library)

One application is to the growth of interest-bearing bank accounts (a topic to which we can all relate, if only in our dreams). The formula for determining compound interest says that if we invest $\$P$ at an annual rate of $r\%$ where interest is compounded k times per year, we end up one year later with a total of

$$\$P\left(1 + \frac{0.01r}{k}\right)^k$$

This is a result bankers know and love.

As an example, suppose we have at our disposal $5,000 to invest in an account offering 10% interest, compounded once at the end of each year. This means that money invested on January 1 and not disturbed will, on December 31, increase in value by 10%. In this case, $P = 5{,}000$, $r = 10$, and $k = 1$ (annual compounding). The formula tells us that at the end of a year our account will be worth:

$$\$P\left(1 + \frac{0.01r}{k}\right)^k = \$5,000\left(1 + \frac{0.01 \times 10}{1}\right)$$

$$= \$5,000(1 + 0.10) = \$5,000(1.10) = \$5,500$$

Fine. But suppose the bank decides to allocate interest differently: instead of giving 10% once a year it gives 5% every six months. This is called semiannual compounding. For the investor, is it an improvement?

In the interest formula, all is the same except now $k = 2$ because we have two interest periods per year. So after a year we end up with an account worth

$$\$P\left(1 + \frac{0.01r}{k}\right)^k = \$5,000\left(1 + \frac{0.01 \times 10}{2}\right)^2 = \$5,000(1.05)^2 = \$5,512.50$$

which is a somewhat better return on investment.

A thought is brewing. If the bank offers more frequent interest payments—such as quarterly, or monthly, or daily—perhaps we will come out even further ahead. To investigate this, we compute the account's value under various interest schemes:

With *quarterly* compounding, we let $k = 4$ and end the year with an account worth

$$\$P\left(1 + \frac{0.01r}{k}\right)^k = \$5,000\left(1 + \frac{0.01 \times 10}{4}\right)^4 = \$5,000(1.025)^4 = \$5,519.06$$

which is better. With *monthly* compounding and $k = 12$, we reach a total of

$$\$P\left(1 + \frac{0.01r}{k}\right)^k = \$5,000\left(1 + \frac{0.01 \times 10}{12}\right)^{12} = \$5,000(1.008333)^{12} = \$5,523.57$$

which is better yet. And with *daily* compounding ($k = 365$), the account grows to

$$\$P\left(1 + \frac{0.01r}{k}\right)^k = \$5,000\left(1 + \frac{0.01 \times 10}{365}\right)^{365} = \$5,000(1.00027397)^{365} = \$5,525.78$$

Drooling with greed, we envision even higher yields if the bank compounds not daily but hourly, or every minute, or every second. In fact, why not imagine the best of all possible interest-bearing accounts: one that compounds interest *continuously*. Then we need not wait even a millisecond for the next interest payment to come along. We imagine the annual 10% rate split among infinitely many compounding periods, each of infinitely short duration. Even as a tree grows, so will our account grow, not in a number of little spurts but in one continuous upward movement.

Formally, to say that interest is compounded continuously means that we let k, the number of compounding periods, tend to infinity. So, after a year of continuous compounding, the account will have grown to:

$$\lim_{k \to \infty} \$P \left(1 + \frac{0.01r}{k}\right)^k = \$P \left[\lim_{k \to \infty} \left(1 + \frac{0.01r}{k}\right)^k\right] = \$Pe^{0.01r}$$

where $0.01r$ is playing the role of x in formula A above. Here, as promised, is the number e in all of its glory.

For our example, the initial investment of $5,000, compounded continuously at 10% for one year, grows to

$$\$5,000e^{0.01 \times 10} = \$5,000e^{0.10} = \$5,000(1.105170918) = \$5,525.85$$

This is the best yield possible for a 10% annual rate.

It should not be surprising that e, useful in determining the continuous increase of a bank account, shows up in other types of continuous growth. For instance populations—be they of bacteria or of humans—can be regarded as increasing continuously, with new individuals born at a rate proportional to the population already at hand. Such a theory was put forth by the British economist Thomas Malthus in 1798 to explain population growth, and half a century later his work would be cited by another scientist, the incomparable Charles Darwin.[2]

Under this simple population model, the number of individuals present at time t, denoted by $P(t)$, is given by

$$P(t) = P_0 e^{rt}$$

where P_0 is the initial size of the population (that is, when we begin observing the situation) and r is a growth-rate constant. Note the similarity to the formula for continuously compounded interest derived above.

As an example, we start with $P_0 = 500$ bacteria in a petrie dish and notice that there are 800 bacteria one hour later. This leads to a growth model in which, after t hours, the population will number $P(t)$ bacteria, where

$$P(t) = 500e^{0.47t}$$

The graph is shown in Figure N.2. Note that for small values of t—that is, for $t = 1, 2,$ and 3—the curve is reasonably flat. The interpretation is that the bacteria population is growing moderately early in the process. But as we move to the right—which is to say, as time passes—the graph starts shooting ever more steeply upward. This reflects a bacteria baby-boom as the germs spill out of the dish, across the desktop, and into the hallway.

To be more precise, at $t = 1$ hour, the formula says we have $P(1) = 500e^{0.47} = 800$ bacteria (which we already knew). After $t = 10$ hours of sustained growth, the formula predicts $P(10) = 500e^{0.47 \times 10} = 500e^{4.7} \approx 55,000$ bacteria, and if breeding continues throughout a 24-hour day, we reach a bacteria population of

$$P(24) = 500e^{0.47 \times 24} = 500e^{11.28} = 39,600,000$$

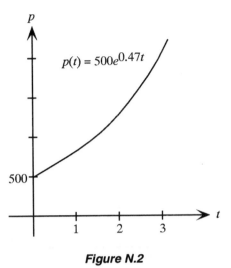

Figure N.2

Were the process to go unchecked for a week, we would have on our hands

$$P\,(168) = 500e^{0.47 \times 168} = 500e^{78.96} \approx$$

$$10{,}000{,}000{,}000{,}000{,}000{,}000{,}000{,}000{,}000{,}000{,}000{,}000{,}000$$

bacteria, which certainly qualifies as an epidemic. These numbers and their sharply rising graph vividly illustrate what is meant when a population is said to be "growing exponentially."

However, it is easy to spot a flaw in this reasoning, for there *must* be an upper bound on the size of any population. Eventually the bacteria will run out of food, or of water, or simply of space. In this sense, unchecked growth is unrealistic growth.

Mathematicians have thus refined their approach to take into account restrictions inherent in population increase. One such refinement, called the **logistic** model, leads to the equation

$$P\,(t) = \frac{Ke^{rt}}{e^{rt} + C}$$

where $P(t)$ is again the population at time t and where K is a number called the **saturation level** that puts a lid on what the environment can support. A logistic growth curve is shown in Figure N.3. For small values of t, the graph resembles that of the previous model. This reflects the observed fact that early growth of a population is pretty much unchecked. But as time passes and we move rightward along the graph, we witness a leveling off in the population as it approaches the horizontal line at $P = K$. This is a graphical reflection of the population approaching its saturation level.

We have, of course, glossed over a multitude of technical points about the origin of these equations. Moreover, biologists have devised even more sophisticated mod-

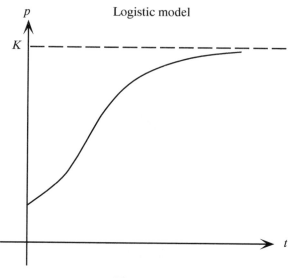

Figure N.3

els to reflect the behavior of populations in nature. (What happens, for instance, if an antibiotic is further restricting bacteria growth?) For our purposes, however, the key point is this: population growth formulas depend on the number e. It arises naturally in describing the biological world around us.

Many other real-life situations lead unexpectedly to this number. Consider the following: A manufacturer of sulfuric acid has a 100-gallon tank filled to the brim with a solution of 25 percent acid and 75 percent water. The manufacturer wants to flush out the tank by introducing a stream of pure water into the top at a rate of 3 gallons per second. To prevent overflow, the mixture is simultaneously pouring out the bottom at the same 3 gallons per second, as depicted in Figure N.4.

It is clear that this process will continuously dilute the contents of the tank. Yet it is also clear that the precise dynamics of this situation are far from simple. It is not as though the incoming water replaces only acid. On the contrary, some of the pure water pumped into the tank is going to be pumped out again as part of the mixture, whereas some of the acid will remain behind in solution. The question confronting mathematicians is to determine the percentage of acid in the tank t seconds after the onset of the flushing process.

An analysis of this problem, employing the techniques of integral calculus, generates the following equation for $P(t)$, the percentage of acid in the tank at any time t:

$$P(t) = \frac{25}{e^{0.03t}}\%$$

What is significant is that, here again, e has floated to the surface.

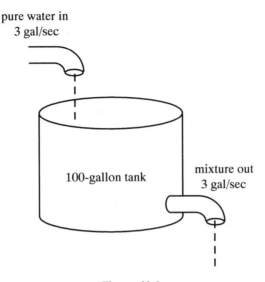

Figure N.4

We take a look at this equation in action. At the outset the tank contained 25 percent acid. After $t = 5$ seconds of pouring in pure water and pumping out the mixture, its contents are diluted to

$$P(5) = \frac{25}{e^{0.03\times5}} \% = \frac{25}{e^{0.15}} \% = 21.52\% \text{ acid}$$

After a minute, the acid percentage is diluted to

$$P(60) = \frac{25}{e^{0.03\times60}} \% = \frac{25}{e^{1.8}} \% = 4.13\%$$

And, if the manufacturer continues the process for a quarter of an hour, so that $t = 15 \times 60 = 900$ seconds, the vat will contain a minuscule

$$P(900) = \frac{25}{e^{0.03\times900}} \% = \frac{25}{e^{27}} \% = 0.000000000047\% \text{ acid}$$

For all intents and purposes, after 15 minutes the vat has been flushed clean.

We encounter e in a different situation if we recall the work of Jakob Bernoulli described in Chapter B. There we saw that the probability of getting exactly 247 heads on 500 flips of a balanced coin is given by the daunting formula

$$\frac{500!}{247! \times 253!} \left(\frac{1}{2}\right)^{247} \left(\frac{1}{2}\right)^{253}$$

The computation of such a probability cannot be undertaken directly. But a bit of mathematical statistics reveals that a close approximation is given by

$$\frac{1}{2\sqrt{250\pi}}\left[\frac{1}{e^{0.025}}+\frac{1}{e^{0.049}}\right]$$

where e again, for a seemingly inexplicable reason, plays a key role (as does π, which seems equally improbable). This expression simplifies to 0.0344, so we have about a 3.44 percent chance of getting 247 heads on 500 flips of a coin. This example illustrates one of the major truisms of probability theory: If a formula is important in the world of statistics, it probably has an e in it.

And so, e is of enormous significance in mathematics, for both theoretical and practical reasons. It is present when we flush vats or flip coins, when we earn interest or watch bacteria grow. Rather like the characters in a Dickens novel, e keeps turning up in the most unexpected places. But whereas the appearances and reappearances of a Dickensian character ask the reader to accept the premise that stupendously unlikely coincidences are commonplace, the appearances and reappearances of e require only that we understand a bit of mathematics.

This, however, is only half the story. Although it is important to find powers of e, it is no less important to be able to reverse the process. Consider the following example. Upon substituting $x = 2$ into formula B, we saw that $e^2 = 7.389056099$. Suppose instead we were asked to determine x knowing that $e^x = 7.389056099$. Of course, this is an easy one: $x = 2$.

But what if we had to find x knowing that $e^x = 5$? We could start guessing various x values, utilize the e^x key on the calculator, and eventually zero in on the answer, but this approach seems a bit roundabout.

To our rescue comes the process of "inverse exponentiation," which undoes whatever e^x does. The function that accomplishes this is called the **natural logarithm,** or more familiarly the natural log, and is denoted in most mathematics texts and calculator keyboards by "ln x." It is without question one of the most important functions in all of mathematics.

For the purposes of this chapter, its one crucial property is the inversion formula

$$\ln(e^x) = x$$

In symbolic form this says what we expressed verbally above: The natural logarithm undoes the exponential. That is, if we start with x, compute e^x, and then insert e^x into the natural log, we return to our starting point x. When $x = 2$, $e^2 = 7.389056099$, and $\ln(e^2) = \ln(7.389056099) = 2$, as the calculator readily confirms. To find x knowing that $e^x = 5$, we take logs of both sides to get

$$\ln(e^x) = \ln 5$$

But the relationship above tells us that $\ln(e^x) = x$, and because $\ln 5 = 1.609437912$, we conclude that

$$x = 1.609437912$$

To summarize: Instead of beginning with x and determining e^x, mathematicians often must go in the other direction, starting with e^x and from this determine x itself. It is in such cases that the natural logarithm earns its keep. Although we shall meet it again in Chapters P and U, for now we illustrate the utility of $\ln x$ with a single example from the shadowy world of crime and punishment, of law and logarithm.

Police were summoned at midnight to a grizzly murder scene where they found the body of Eddie the Weasel, a notorious criminal with reputed underworld connections. Upon arrival, officers noted that the air temperature was a mild 68°F and the body temperature was 85°. At 2:00 A.M., after fingerprints had been taken and suspects questioned, the body had further cooled to 74°.

Acting on a tip, the police arrest Clare Voyant, Eddie's visionary girlfriend. Clare had spent the evening in Louie's Bar, drinking a bit too much and threatening Eddie's life. She stormed out at 11:15 P.M. in a foul mood. It looked like an open-and-shut case.

Fortunately, Clare knew natural logarithms. She also knew Newton's law of cooling, a cornerstone of the theory of heat dissipation. Newton's law says that the rate at which an object cools is proportional to the difference between the object's temperature and the temperature of its surroundings. In everyday language this means that when an object is much hotter than the outside air, its rate of cooling is high, so it cools down very fast; when a body is just a little hotter than its surroundings, its rate of cooling is low and it cools slowly.

Newton's law applies to anything that is cooling off, be it a hot potato just out of the oven or a lifeless corpse lying upon the sidewalk. A *living* person is not cooling off. Human metabolism insures that body temperature is maintained at or about 98.6°F. But a nonliving person ceases to generate heat and thus cools, potato-like, according to Newton's law.

Upon translating the verbal description above into a concise mathematical formula and applying calculus, Clare derived the following equation for T, the temperature of the body t hours after midnight:

$$T = 68° + \frac{17°}{e^{0.5207t}}$$

Note again the presence of the number e. With a calculator one can check that at midnight, when $t = 0$, the body temperature was

$$T = 68° + \frac{17°}{e^{0.5207 \times 0}} = 68° + \frac{17°}{1} = 68° + 17° = 85°\text{F}$$

as the police determined upon arrival. Likewise, at 2:00 A.M., when $t = 2$, the formula gives a body temperature of

$$T = 68° + \frac{17°}{e^{0.5207 \times 2}} = 68° + \frac{17°}{2.8349} = 68° + 6.000° = 74°\text{F}$$

which again reflects the police observation. In other words, the formula works fine at those two times for which we have actual data.

But the central challenge for Clare was to determine *when* Eddie the Weasel met his end. She somehow had to use this formula to reverse the cooling process and thereby calculate the time t when Eddie's body temperature last stood at a normal 98.6°F. This, of course, was his time of death. From that point onward, the deceased Eddie was merely cooling his heels (and everything else).

So, we substitute the normal body temperature of $T = 98.6°$ into the cooling equation to get

$$98.6° = 68° + \frac{17°}{e^{0.5207t}}$$

Subtracting 68° from each side and cross-multiplying gives $(30.6°)e^{0.5207t} = 17°$, and then dividing both sides by 30.6° brings us to

$$e^{0.5207t} = \frac{17°}{30.6°} = 0.5555$$

The object was to find t. To do this, Clare took logs of both sides of the equation:

$$\ln(e^{0.5207t}) = \ln(0.5555)$$

Now, $\ln(0.5555) = -0.5878$ and the inversion formula above guaranteed that $\ln(e^{0.5207t}) = 0.5207t$. Consequently,

$$0.5207t = \ln(e^{0.5207t}) = \ln(0.5555) = -0.5878$$

Therefore at time $t = -0.5878/0.5207 = -1.13$ hours, Eddie's body temperature was 98.6°F.

Here t, measured as the number of hours *after* midnight, was negative. The interpretation is immediate: The body had 98.6° temperature 1.13 hours *before* midnight. In other words, Eddie the Weasel starting cooling off—which is to say, died—about 68 minutes before 12:00 A.M. This put his time of death at 10:52 P.M. But at that moment Clare was known to be drinking in Louie's Bar. She had an ironclad alibi!

At the trial, Clare's lawyer introduced the evidence above, appealed eloquently to "the laws of nature and nature's log," and won easy acquittal before a jury of mathematically sophisticated peers. Thanks to logarithms, justice was served.

Forensic pathologists certainly know about natural logarithms. So do geneticists, geologists, and virtually everyone else who studies dynamic real-world phenomena. Intuition aside, it is a remarkably important, pervasive, and useful idea. We trust that the jury of readers, upon considering the evidence above, will find the number e and its counterpart, the natural logarithm, not guilty of gross insignificance.

rigins

Of the earliest beginnings of mathematics there are no lasting traces. Such information is irretrievably lost, and just as we cannot determine who uttered the first word or sang the first song, we have no idea who discovered the first mathematic.

We do know that the rudiments of arithmetic and geometry go back a very long way. Before written history, before writing itself, humans had developed some concept of "multitude" or "number," and there are artifacts to support this. A bone from Africa exhibiting what can only be interpreted as tally marks is at least 10,000 years old.[1] In this prehistoric time, our ancestors were counting *something*, and the scratching of marks into bone provided them—and us—with a permanent record of their counts. It may have had a modest beginning, but mathematics was on its way.

Clearly the subject did not arise in a single spot, any more than storytelling or music or art had a unique birthplace. Mathematical concepts appear in the historical record from many different regions around the globe and, as we saw when discussing the Pythagorean theorem in Chapter H, the same principle may be discovered in more than one place. This suggests not only the universality of mathematics but also the universal tendency of humans to mathematize.

In this chapter we take a brief look at a few early mathematical landmarks. Somewhat arbitrarily, our survey will be restricted to the period before A.D. 1300 and to discoveries from Egypt, Mesopotamia, China, and India—four cultures that have served as pillars of human civilization.

Egyptian mathematics can be traced back at least 4,000 years before disappearing into prehistory. Scholars have deciphered papyrus rolls predating 1500 B.C., some of which are indisputably mathematical. Perhaps the most famous is the

Ahmes papyrus from about 1650 B.C. and named for the scribe who wrote it. This 18-foot-long document was purchased in Egypt in 1858 and now resides somewhere in the British Museum. In it, the scribe Ahmes promised "insight into all that exists, knowledge of all obscure secrets."[2] Although the papyrus fell well short of this ambitious pledge, he did bequeath to future generations some fascinating glimpses of Egyptian arithmetic and geometry.

The Ahmes papyrus contains dozens of problems and their accompanying solutions. Often these are what we would today call "story problems," exhibiting much the same flavor (and the same artificiality) as their modern counterparts. For instance, the 64th problem of the Ahmes papyrus is:

> Divide 10 hekats of barley among 10 men so that the common difference is 1/8 a hekat of barley.[3]

Those well grounded in algebra are apt to introduce x as the number of hekats given to the first man. Then the next man gets $x + 1/8$, the third gets $x + 2/8$, and so on up to the tenth man's allocation of $x + 9/8$. Because altogether there are ten hekats of barley to be distributed, we arrive at the equation

$$x + \left(x + \frac{1}{8}\right) + \left(x + \frac{2}{8}\right) + \left(x + \frac{3}{8}\right) + \left(x + \frac{4}{8}\right) + \left(x + \frac{5}{8}\right) + \left(x + \frac{6}{8}\right)$$
$$+ \left(x + \frac{7}{8}\right) + \left(x + \frac{8}{8}\right) + \left(x + \frac{9}{8}\right) = 10$$

Algebraically, this simplifies to $10x + 45/8 = 10$ and so

$$x = \frac{1}{10} \times \left(10 - \frac{45}{8}\right) = \frac{7}{16} \text{ hekats of barley}$$

as the first individual's share. The next person gets

$$\frac{7}{16} + \frac{1}{8} = \frac{9}{16} \text{ hekats}$$

the next gets

$$\frac{9}{16} + \frac{1}{8} = \frac{11}{16} \text{ hekats}$$

and so on.

It must be emphasized that the Egyptian solution did not have this explicit algebraic character, for symbolic algebra lay thousands of years in the future. Nonetheless, Ahmes gave the solution correctly, stating that the first individual should get

$$\frac{1}{4} + \frac{1}{8} + \frac{1}{16} \text{ hekats of barley}$$

As significant as the right answer is the way it was expressed: as the sum of fractions each with numerator of 1. These are called *unit fractions,* and the Egyptians used them almost exclusively. Thus Ahmes' answer to the barley problem consisted of the sum of three unit fractions rather than as the equivalent 7/16. To the modern mind, this is both peculiar and unnecessarily complicated.

But this practice fit nicely with the Egyptian notational scheme: to represent a reciprocal, they used a symbol—looking something like a floating cigar—atop an integer. A modern counterpart would be to let $\overline{2}$ stand for 1/2 or $\overline{7}$ stand for 1/7. The number of hekats in the problem above could thus be concisely written as $\overline{4} + \overline{8} + \overline{16}$. Such a notation is simple but obviously requires numerators of 1. For the Egyptians, all fractions had to be assembled from unit fractions, with the lone exception of 2/3, which had its own unique symbol.

Ahmes provided an extensive list of unit fraction representations like that above. Such a list must have served the Egyptians much as logarithmic or trigonometric tables served mathematicians in the days before calculators. All in all, the Egyptians proceeded quite nicely with their unit fraction method, cumbersome though it seems to us today.

But the Egyptian mathematical contribution was not limited to arithmetic/algebraic problems like that above. Ahmes' papyrus, for instance, contains a selection of geometric questions, perhaps the most intriguing of which is problem 50:

A circular field has diameter 9 khet. What is its area?[4]

According to the scribe, the answer is obtained by subtracting one-ninth of the diameter and then squaring the result. For this problem, with diameter $D = 9$, the circular area is

$$\left[D - \frac{1}{9}D \right]^2 = \left[9 - \frac{1}{9}(9) \right]^2 = 8^2 = 64$$

What is interesting is that an estimate of π can be unearthed from this solution. Translated into modern notation, Ahmes said that a circle with diameter D has area

$$\left(D - \frac{1}{9}D \right)^2 = \left(\frac{8}{9}D \right)^2 = \frac{64}{81}D^2$$

Because the true area of a circle is

$$\pi r^2 = \pi \left(\frac{D}{2} \right)^2 = \frac{\pi}{4}D^2$$

the Egyptian result amounts to

$$\frac{64}{81}D^2 = \frac{\pi}{4}D^2$$

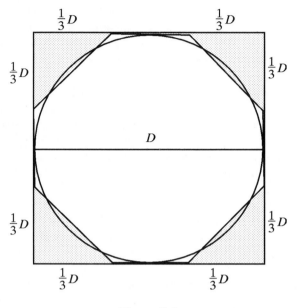

Figure O.1

From this it follows that

$$\pi = \frac{4 \times 64}{81} = \frac{256}{81} = \left(\frac{16}{9}\right)^2 = \left(\frac{4}{3}\right)^4 \approx 3.1605$$

which is often cited as the Egyptian approximation of π. Given its antiquity, the accuracy is impressive. But how did they come up with it?

Although no one is certain, a possibility is that the circular area was replaced by that of a related octagon, as indicated in Figure O.1. About a circle of diameter D circumscribe a square and remove from the square's corners the four shaded isosceles triangles with equal sides of length $(1/3)D$. What remains is an octagon whose area approximates that of the original circle. Because each missing triangle has area

$$\frac{1}{2}(\text{base} \times \text{height}) = \frac{1}{2}\left(\frac{D}{3}\right) \times \left(\frac{D}{3}\right) = \frac{1}{18}D^2$$

we see that

area(circle) $\approx$ area(octagon)

$= $ area(square) $- 4$ area(isosceles triangle)

$$= D^2 - 4\left(\frac{1}{18}D^2\right) = D^2 - \frac{2}{9}D^2 = \frac{7}{9}D^2 = \frac{63}{81}D^2$$

This is very close to Ahmes' value of $(64/81)D^2$ and suggests that the octagonal approximation to circular area may have led to the Egyptian formula.

There are those who dismiss this explanation.[5] But the technique of approximating circles with polygons is *exactly* what Archimedes employed a millennium and a half later to come up with his significantly sharper estimate of $\pi \approx 3.14$. As we saw in Chapter C, his improved accuracy resulted from approximating the circle by regular 96-sided polygons rather than the fairly crude octagon above. It is possible that Archimedes owed a debt to his Egyptian predecessors.

If the Nile Valley was one source of early mathematical ideas, so, too, was the Tigris-Euphrates region of Mesopotamia. Mesopotamian political history was far more tumultuous than that of Egypt, for the area was the scene of conquests and reconquests by many tribes and factions. Although such power transfers make the designation imprecise, the work we are about to discuss is usually called "Babylonian mathematics."

The golden age of Babylonian scholarship came at the time of Hammurabi (ca. 1750 B.C.), roughly contemporary with Ahmes in Egypt. Fortunately for later scholars the Babylonians wrote on clay tablets rather than on papyrus. The good fortune is due to the fact that papyrus deteriorates over time, whereas the heavy tablets, which were inscribed and then baked, are far less biodegradable. We thus have thousands of relics, both complete and in shards, and among these is a good sampler of their mathematics.

If unit fractions are an identifying characteristic of Egyptian mathematics, the base-60 system of enumeration is its counterpart in the Babylonian. Historians have long expressed both admiration and surprise that the Babylonians adopted such a scheme. They used two symbols: one, looking a bit like a T, stood for 1 and the other, resembling our $<$, represented 10.

For small numbers, their notation was unexceptional: 2 was written TT; 12 was $<$ TT; 42 was

$$\begin{matrix} << \\ << \end{matrix} \text{TT}$$

and so on. But when the Babylonians reached 60, their symbols took on a different meaning. If regarded as occupying a *position*, T could mean not one unit but one 60. If occupying a different position, it could be one $60^2 = 3,600$. Consequently, the number 82 appears as T $<<$ TT, meaning one 60 plus two 10s plus two 1s. Note that the symbol T here means different things depending on its location in the numeral.

This innovation not only streamlined the process of representing numbers but kept the collection of primitive symbols to a minimum. By contrast the Roman numeral system, which was not a place-system, required a bevy of symbols such as I, V, X, L, C, D, M. As numbers rose into the millions, billions, and trillions, more Roman letters would have to be introduced until, conceivably, their alphabet would be exhausted. (Of course, the Romans had no need for numbers of such size, this being millennia before the arrival of the national debt.)

Equally significant, Babylonian notation allowed them to represent fractions with ease. One ancient tablet introduced the number whose square is two—we would write $\sqrt{2}$ —as

$$\text{T} \ll \frac{\text{TT}}{\text{TT}} \underset{\ll}{\overset{\ll}{<}} \text{T} \quad <$$

In our notation, the groupings reduce to the string of integers 1-24-51-10. The Babylonians had nothing corresponding to a decimal point, so we are left to determine the integer and fractional parts of this expression. But because $\sqrt{2}$ is a bit greater than 1, it is clear that the first number is the integer and the others represent fractions.

But which fractions? In our base-10 system, the digit immediately to the right of the decimal point is the number of tenths; the next digit is the number of hundredths; the next, the number of thousandths; and so on. In precisely this manner, we interpret the expression following the 1 as the number of sixtieths; the next symbol is the number of $1/60^2$ or 3,600ths; and the last is the number of $1/60^3$ or 216,000ths. Consequently, the Babylonian 1-24-51-10 amounts to our

$$1 + \frac{24}{60} + \frac{51}{3,600} + \frac{10}{216,000} = 1 + 0.4 + 0.014167 + 0.000046 = 1.414213$$

As we saw in Chapter K, a nine-place estimate of $\sqrt{2}$ is 1.414213562, so the Babylonian estimate was impressive. They evidently had mastered base-60 arithmetic.

From a modern viewpoint, their system has one striking omission: The Babylonians had no symbol for zero. This omission could lead to misinterpretations because < T could mean 11, or $10 \times 60 + 1 = 601$, or $10 \times 3,600 + 1 = 36,001$, or $10 \times 3,600 + 1 \times 60 = 36,060$. As with fractions, much confusion could be eliminated by considering the context, but the introduction of a zero as "place holder" was necessary to relieve ambiguity.

Interestingly, the Babylonians never quite took such a step. By the Seleucid period in middle of the first millennium B.C., an *internal* place-holder notation was introduced that made it possible to distinguish, for instance, between 61 and 601. But they did not address the issue of zeros at the end of a number, so their notation never distinguished 620 from 62,000. A true zero finally appeared, centuries later, in the mathematics of India and, independently, in that of the Mayans of Central America. When it came, it was a great innovation.

But another question arises: *Why* did the Babylonians choose 60 as their base? Anthropological and archaeological studies of human culture have found the common number bases to be 2, 5, 10, and somewhat less frequently 20. These correlate nicely with features of the human anatomy: arms, fingers on one hand, fingers on both hands, fingers and toes. Put another way, people had bodily references in case their arithmetic failed them.

But why 60? Although it is impossible to answer this question with certainty, it seems suggestive that a year contains almost exactly $6 \times 60 = 360$ days. No one studying the origins of mathematics can fail to recognize the impact of astronomy, and no astronomical measurement is more critical than the length of the year. Per-

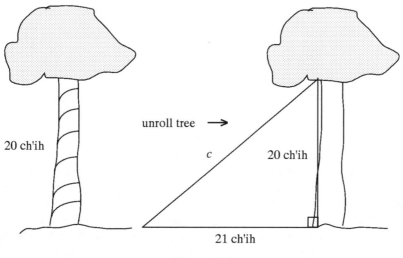

unroll tree →

20 ch'ih

20 ch'ih

c

21 ch'ih

Figure O.2

haps the year of (roughly) 360 days was decisive in elevating the number 60 to a key position in Babylonian arithmetic. In any case, its presence has come down to us in such base-60 measurements as 60 seconds in a minute, 60 minutes in an hour, and 360 degrees in a circle.

Babylonian mathematics at least equalled that of the Egyptians in laying a foundation for later discoveries in the eastern Mediterranean. But this was not the only place that mathematics flourished 2,000 years ago. At the other end of Asia the Chinese had established their own impressive mathematical tradition.

We have already encountered Chinese mathematics in Chapter H, where we looked at the proof of the Pythagorean theorem as found in the *Chou Pei Suan Ching*. The Chinese clearly understood the theorem in its greater generality, as is evident from a collection of problems that used it. For instance, the following appears in the *Chiu chang suan-shu*, a mathematical treatise dating back at least two millennia and sometimes called the Chinese equivalent of Euclid's *Elements*. The fifth problem of the last of nine chapters of the *Chiu chang* is:

> A tree of height 20 ch'ih has a circumference of 3 ch'ih. There is an arrow root vine which winds seven times around the tree and reaches the top. What is the length of the vine?[6]

As shown at the left in Figure O.2, the vine ascends in a helix-shaped curve around the cylindrical tree. The objective is to find its length. To do so, imagine this scenario: Let the foot of the vine be fixed to the ground and the tree then be "rolled" toward the right through seven revolutions. As the tree moves, the vine unwinds until we reach a final configuration with the vine stretched tightly between treetop and ground, as shown on the right in the same figure.

4	9	2
3	5	7
8	1	6

Figure O.3

This produces a right triangle whose height is 20, the height of the tree and whose width is the distance covered by a point at the base of the trunk as the tree rolls along. For each revolution, the point moves through one circumference of the tree—3 ch'ih—and therefore the triangle's base is $7 \times 3 = 21$ ch'ih. So we have a right triangle with legs 20 and 21 and with hypotenuse c, the vine's length. By the Pythagorean theorem, $c^2 = a^2 + b^2 = 20^2 + 21^2 = 400 + 441 = 841$, and so $c = \sqrt{841} = 29$ ch'ih. This is the answer given in the *Chiu chang* by the Chinese masters of 2,000 years ago. It is a bit sobering to speculate how today's populace would fare on the vine problem.

Moving from geometry to arithmetic, the Chinese were fascinated by *magic squares,* which are square arrays of whole numbers each of whose rows, columns, and main diagonals has the same sum. As with all ancient mathematics, it is difficult to assign a precise date to the origins of the subject, but there is a legend that Emperor Yu of 5,000 years ago copied a magic square from the shell of a mystical turtle. Although mathematicians prefer attributing results to pure reason rather than magical reptiles, there is no question that the Chinese were the ancient masters of such numeric arrangements.

In Figure O.3 we see a 3×3 square containing the whole numbers from 1 to $3^2 = 9$. Note that each row, each column, and the two main diagonals sum to 15. This 3×3 magic square, called the *Lho shu*, carried a special meaning to the Chinese because of the harmony and balance, the yin and yang, that gave the mathematical configuration an almost spiritual dimension.

This square is not too hard to devise, but what about its 4×4 or 5×5 counterparts? Their construction is a bit trickier and requires a more subtle theory, one that intrigued not only the Chinese but later the Arabs and still later Benjamin Franklin, who concocted magic squares when political debates became tedious.

If we wish to create an $m \times m$ magic square, the first step is to ascertain the common sum of each row, column, and main diagonal. Because we must distribute about the square all the numbers from 1 to m^2, we know the total of *all* integers in the square is $1 + 2 + 3 + \ldots + m^2$. As we saw in Chapter J, the sum of the first n whole numbers is given by the simple formula

$$\frac{n(n + 1)}{2}$$

Thus, regardless of their arrangement, the total of *all* entries in an $m \times m$ magic square will be

$$1 + 2 + 3 + \ldots + m^2 = \frac{m^2(m^2 + 1)}{2}$$

Of course, this same sum will be obtained by adding the totals of each of the square's m rows. Because each row sum is the same, each must be $1/m$th of the total, and thus the sum of each row (or each column) of an $m \times m$ magic square is

$$\frac{1}{m} \times \frac{m^2(m^2 + 1)}{2} = \frac{m(m^2 + 1)}{2}$$

For instance, if we wish to build a 5×5 magic square, each row, column, and main diagonal must sum to

$$\frac{5(5^2 + 1)}{2} = 65$$

Much work remains in order to arrange the numbers magically, but this preliminary calculation lets us know what row and column totals to shoot for. The Chinese were perfectly capable of completing the problem, as is evident from a 5×5 magic square shown in Figure O.4 and attributed to Yang Hui of the thirteenth century.

Notice that, as predicted, each row, column, and main diagonal sums to 65. Other internal patterns are evident. For instance, if we extract the 3×3 square surrounding the central entry of 13 (see Figure O.5), we find it is a modified magic square—using the numbers 7, 8, 9, 12, 13, 14, 17, 18, and 19—with each row, column, and diagonal summing to 39. This and other examples of "order within order" held special appeal to those seeking divine perfection in a numerical array.

1	23	16	4	21
15	14	7	18	11
24	17	13	9	2
20	8	19	12	6
5	3	10	22	25

Figure O.4

14	7	18
17	13	9
8	19	12

Figure O.5

Leaving the Chinese, we must hurry to mention one additional civilization whose contributions were of the highest significance: the Hindu culture of India. Indian mathematics dates roughly to the time of the Egyptian papyri and the Babylonian tablets, and a fascinating but unresolved question is the extent of contact that existed among these peoples. There are certainly grounds for suspecting an Indian-Chinese mathematical interaction, but the jury is still out as to its magnitude and direction.

All that aside, the Indians excelled in mathematics. Among their foremost achievements was the development of trigonometry. Much of their work in this area passed through later Arab cultures and into Europe of the fifteenth century. The modern world owes much to the great Indian trigonometers.

The Indians also solved sophisticated problems of an algebraic, albeit non-symbolic, nature. An example of such a problem is due to Bhaskara—also known as Bhaskaracharya or "Bhaskara the Teacher"—who lived around A.D. 1150. For instance, one challenge was to determine whole numbers such that 61 times the square of the first is one less than the square of the second. In modern notation, this amounts to finding x and y so that $61x^2 = y^2 - 1$. The problem, which Bhaskara correctly solved, turned up again in seventeenth-century Europe and gave mathematicians quite a workout. This is hardly surprising given his answer: $x = 226,153,980$ and $y = 1,766,319,049$.[7]

The Indians also left us some intriguing geometrical results, one of the most spectacular of which is known as Brahmagupta's formula for the area of cyclic quadrilaterals. A *cyclic quadrilateral* is a four sided figure that can be inscribed in a circle, as shown in Figure O.6. Brahmagupta, an astronomer/mathematician from the seventh century A.D., stated that the *area* of any such quadrilateral with sides a, b, c, and d is given by

$$\sqrt{(s - a)(s - b)(s - c)(s - d)}$$

where

$$s = \frac{1}{2}(a + b + c + d)$$

is called the quadrilateral's *semiperimeter.*

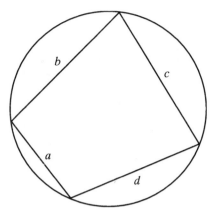

Figure O.6

To see it in action, consider the rectangle with sides a and b in Figure O.7. A circle, of course, can be circumscribed about the rectangle by placing the circle's center at O, the intersection of the rectangle's diagonals. Because a rectangle is thus a cyclic quadrilateral, we can apply Brahmagupta's formula. In this case,

$$s = \frac{1}{2}(a + b + a + b) = \frac{1}{2}(2a + 2b) = a + b$$

so $s - a = (a + b) - a = b$ and $s - b = (a + b) - b = a$. Hence the area of the rectangle is

$$\sqrt{(s - a)(s - b)(s - a)(s - b)} = \sqrt{b \times a \times b \times a} = \sqrt{a^2 b^2} = ab$$

Of course, we hardly need a weapon as powerful as Brahmagupta's formula to discover that the area of a rectangle is the product of its base and height. This is rather like mowing the lawn with a combine.

But the next example, taken from an ancient Indian text, is less elementary.[8] Here we seek the area of the cyclic quadrilateral with sides $a = 39$, $b = 60$, $c = 52$, and $d = 25$, as shown in Figure O.8. Without the assistance of Brahmagupta, this would present substantial difficulties; with Brahmagupta, the answer is immediate. Here the quadrilateral's semiperimeter is

$$s = \frac{1}{2}(39 + 60 + 52 + 25) = 88$$

and its area is $\sqrt{(88 - 39)(88 - 60)(88 - 52)(88 - 25)} = 1{,}764$.

Brahmagupta's formula has an interesting consequence. If in Figure O.9 we slide vertex D along the circle to C, the inscribed quadrilateral is transformed into triangle

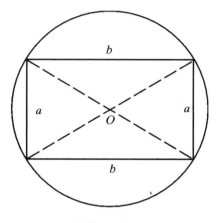

Figure O.7

ABC. Under this transformation, the length $\overline{CD} \to 0$, so the triangle, viewed as a "degenerate" quadrilateral, has area

$$\sqrt{(s-a)\,(s-b)\,(s-c)\,(s-0)} = \sqrt{s(s-a)\,(s-b)\,(s-c)}$$

where s is now the semiperimeter of $\triangle ABC$. Some readers may recognize this as Heron's formula for triangular area, named after the Greek mathematician who furnished a clever proof around A.D. 75. In this sense Brahmagupta's formula provides an extension of Heron's formula to cyclic quadrilaterals. It is a striking piece of geometry.

We have briefly mentioned one of the greatest achievements of the Indian mathematicians: the introduction of zero into a base-10 number system. The origins of this idea are impossible to date with precision, but it may go back to the middle of the

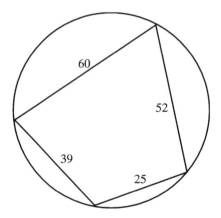

Figure O.8

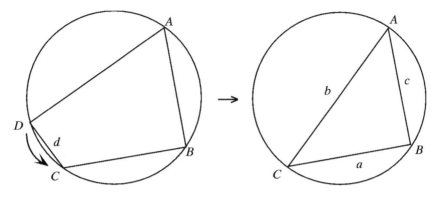

Figure O.9

first millennium A.D. Documents and inscriptions from this period show the zero very clearly—looking much as it does today. This innovation proved useful, not only as a theoretical construct but also as a computational device. So adept were the Indians at using their numeral system, including zero, that their techniques were quickly adopted by the Arabs with whom they came in contact. By the end of the first millennium, Arabic scholars were writing books about the wonderful "Indian arithmetic."

It was through the Arabs that these ideas eventually filtered westward into Europe. A critical step came with the publication of the *Liber abaci* by Leonardo of Pisa in 1202. Leonardo, who is today better known by the name Fibonacci, spent much of his youth in Northern Africa where he learned the Arabic language and studied Islamic mathematics. In this way he mastered what has come to be called the "Hindu-Arabic" numeration system. Fibonacci's book brought these ideas to the Italian centers of learning, from which they subsequently spread across the European continent.

The story of zero typifies much of the history of mathematics. An idea is born; it is refined and transmitted over the miles and over the centuries; it becomes a part of the multinational mathematical culture. Mathematics is a creation in which all the world can proudly share.

Or so it should be. However, the question of mathematical origins has recently become a small skirmish in the larger battle challenging the traditional view of Western civilization. One finds two extreme positions. The first, Eurocentrism, holds that true mathematics started in Greece, where it sprang to life from the minds of brilliant, indigenous thinkers. This school barely credits mathematical achievement or influence from any other quarter.

Opposed to this view are the multiculturalists, who assert that mathematics owes its origins equally to many different peoples around the globe. This position attributes a wide array of discoveries to non-European civilizations and asserts that Eurocentric scholars have tried to distort history in order to enhance their own national, religious, or racial identity.

Needless to say, such arguments quickly grow heated even within the usually sedate halls of mathematics. As with most scholarly battles, extremists tend to monopolize the spotlight while the truth lies somewhere in between.

No one, even those most repelled by the excesses and errors of European civilization, can ignore the colossal achievements of mathematicians from that part of the world. It cannot be denied that the foremost contribution of the Greeks—imparting to mathematics the emphasis on logical proof that so fundamentally characterizes it to this day—brought to the subject a level of sophistication and abstraction missing from earlier traditions. Scholars have uncovered nothing as sophisticated as Euclid's *Elements* or Archimedes' *On the Sphere and the Cylinder* from any equally remote mathematical tradition. Greek mathematics was something new and remarkable. And certainly the subsequent discoveries of Newton, Leibniz, or Euler are no less magnificent because these individuals happened to hail from England, Germany, or Switzerland.

But it is equally clear that Greek mathematics did not arise in a vacuum. Other cultures, both before and after the Greeks, have contributed much to the mathematical harvest. All too often this fact has been overlooked by scholars who behaved as though sharing credit was somehow equivalent to forfeiting it. Those who contend that the Eurocentric interpretation of mathematics history has been terribly myopic can back up their claims with an impressive list of evidence.

One such piece of evidence that tends to be overlooked by the more ardent Eurocentrists comes from the Greeks themselves. The historian Herodotus of the fifth century B.C. wrote that the Egyptian interest in area measurements was "the way that geometry was discovered and came to be introduced among the Greeks."[9] Contact certainly existed between Egypt in Africa and Greece in southern Europe, and initially the flow of ideas was northward. Without in any way detracting from the Greek achievement, it would be highly misleading not to recognize the Egyptians of an earlier time who significantly affected the course of Greek thought.

The Babylonian contributions to arithmetic and enumeration, the Chinese discoveries about the theory of numbers, the Indian trigonometry, the later Arabic algebra—all of these have been blended into the feast of modern mathematics. Removing any from the menu would significantly diminish the whole.

As with so many issues from distant past, the question of origins remains without a final answer. Better scholarship, more extensive translations, and a lucky archaeological discovery or two may provide us with clearer glimpses into the beginnings of mathematics. But much of today's rhetoric over who first discovered what harks back to an earlier episode: the calculus priority dispute described in Chapter K. Like that unfortunate controversy—an explosive mixture of truth, falsehood, and national pride—the current debate can quickly lose sight of an essential fact about the origin of mathematical ideas: In this sphere of human achievement, there is plenty of glory to go around.

Prime Number Theorem

$\pi(\boldsymbol{x})/\boldsymbol{x} \approx 1/\ln x$

Here we return to our haunts of Chapter A: the properties of prime numbers. As we noted, these serve as unique building blocks for the system of natural numbers and thus have received, and deserved, considerable attention over the centuries.

Among the many questions involving primes, one of the most interesting concerns their distribution among the integers. Are they spread among their non-prime relatives in a purely random fashion? Or is there some rule, some discernible pattern with which the primes occur? The answer to the latter question is "sort of." If this seems like an evasive and unsatisfactory response, the present chapter hopes to demonstrate that it really is a fairly bold one. It paraphrases of one of the most spectacular results in all of mathematics, the prime number theorem.

Anyone investigating the distribution of primes should begin with a list. Below are recorded the 25 primes less than 100:

$$2, 3, 5, 7, 11, 13, 17, 19, 23, 29, 31, 37, 41,$$
$$43, 47, 53, 59, 61, 67, 71, 73, 79, 83, 89, 97$$

If there is a pattern here, it is far from obvious. Of course, all primes after 2 are odd, but this is not very helpful. We notice a few gaps in the primes: There are none from 24 to 28 nor from 90 to 96, the latter being a run of seven consecutive composites. On the other hand we see that some primes occur only two units apart—such as 5 and 7 or 59 and 61. Such back-to-back primes, those of the form p and $p + 2$, are called ***twin primes.***

To increase the amount of data, we collect all the primes from the second hundred numbers; that is, from 101 and 200:

$$101, 103, 107, 109, 113, 127, 131, 137, 139, 149, 151,$$

$$157, 163, 167, 173, 179, 181, 191, 193, 197, 199$$

This time there are 21 such numbers. Again we see gaps, such as the nine straight composites between 182 and 190, yet prime twins persist right up to 197 and 199.

In a study of the overall distribution of primes, it would seem that gaps (in which consecutive primes are quite far apart) and twins (in which consecutive primes are quite close together) should play an important role. Are there longer gaps among the primes? Is the supply of twin primes infinite? Interestingly, the first of these questions is easily answered but the second is among the unsolved mysteries of number theory.

Start with the easy one. Suppose we are asked to generate a string of five consecutive composites. Using the factorial notation from Chapter B, we consider the numbers

$$6! + 2 = 722, 6! + 3 = 723, 6! + 4 = 724, 6! + 5 = 725, 6! + 6 = 726$$

It is easy to see that none of these is prime, but it is more instructive to ask *why* this is the case. The first number is $6! + 2 = 6 \times 5 \times 4 \times 3 \times 2 \times 1 + 2$. Because 2 is a factor of both $6!$ and of itself, 2 is a factor of the sum $6! + 2$. Hence $6! + 2$ is not a prime. But neither is $6! + 3 = 6 \times 5 \times 4 \times 3 \times 2 \times 1 + 3$, as 3 divides evenly into both terms and consequently into the sum. Likewise, 4 is a factor of both $6!$ and 4 and thus of their sum, 5 is a factor of $6! + 5$, and 6 is a factor of $6! + 6$. Because each of these numbers has a factor, none is a prime. We have thereby generated five consecutive nonprimes.

It could be convincingly argued that we made the search much too complicated. After all, the five straight composites 24, 25, 26, 27, and 28 would serve just as well. Why introduce factorials that carry us up into the 700s?

The answer is that we need a general procedure. If we were asked to find a string of 500 straight composites, scanning of a list of primes would be unrealistic, but the reasoning used above will provide such a string in exactly the same manner.

That is, begin with the number $501! + 2$ and take the integers from there all the way to $501! + 501$. It is evident that this gives 500 consecutive whole numbers. Almost as evident is that all are composite, for 2 divides evenly into $501! + 2$, 3 divides into $501! + 3$, and so on all the way to 501, which divides evenly into $501! + 501$. Here are 500 consecutive composites.

The very same procedure starting with $5,000,001! + 2$ would yield five *million* consecutive numbers with no prime among them, and we could just as easily generate five billion or five trillion consecutive composites. This argument has a stunning consequence: There exist arbitrarily long gaps among the primes.

This means that if we continued as above in counting the primes among every hundred integers, we would reach a point at which there would be none at all—a hundred straight numbers devoid of primes. But the situation is even stranger. When we come to the string of five million consecutive composites, we would examine

50,000 consecutive groups of a hundred integers each and never find a prime among them! At that point it would seem virtually certain that we have run out of primes altogether.

Anyone believing that is referred back to the proof of the infinitude of primes in Chapter A. Enormous gaps there may be, gaps so large that no human could count them in a lifetime, yet beyond these somewhere must lie more primes, always more primes. They are literally inexhaustible.

What about the other issue? Is the supply of twin primes similarly inexhaustible? Number theorists have struggled with this problem for centuries. Even among very large numbers, twins pop up now and then. The primes 1,000,000,000,061 and 1,000,000,000,063 are an example. But to this day no one can prove there are infinitely many prime twins. The question remains unresolved.

Although this problem continues to baffle the finest mathematical minds, the question of the infinitude of *triplet primes* is easy to settle. We shall say that three primes are triplets if they take the form p, $p + 2$, and $p + 4$. For instance, the primes 3, 5, and 7 are triplets. Are there infinitely many sets of these?

To answer this question, we first observe that when any number is divided by 3, the remainder must be either 0, 1, or 2. So if we have prime triplets p, $p + 2$, and $p + 4$ and divide p by 3, there are three possible outcomes.

Perhaps the remainder is zero. That is, p might be a multiple of 3 or, in more symbolic form, $p = 3k$ for some whole number k. If $k = 1$, then $p = 3$, and we have stumbled upon the triplets 3, 5, and 7 again. But if $k \geq 2$, then $p = 3k$ is not prime because it would have two proper factors, 3 and k. It follows that 3, 5, and 7 are the *only* possible triplets under this case.

Alternately, the remainder upon dividing p by 3 might be 1 so that $p = 3k + 1$ for some integer $k \geq 1$. (Note that we can eliminate $k = 0$, because $p = 3(0) + 1 = 1$ is not a prime.) For this case the second member of the triplet is $p + 2 = (3k + 1) + 2 = 3k + 3 = 3(k + 1)$. Obviously $p + 2$, having factors 3 and $k + 1$, cannot be a prime. We conclude that there are no prime triplets for this case.

Finally, suppose p divided by 3 leaves a remainder of 2. Then $p = 3k + 2$ for some integer $k \geq 0$. Hence the third of the triplets is $p + 4 = (3k + 2) + 4 = 3k + 6 = 3(k + 2)$. But then $p + 4$ is not a prime because it has a factor 3. No prime triplets fit into this category either.

Collecting our results, we see that the lone set of prime triplets is the easy one: 3, 5, and 7. The answer to the question "Are there infinitely many prime triplets?" is therefore a resounding "No." There is only one. Yet replacing the word *triplet* by *twin* converts this into a world-class problem. What a difference a word makes.

All of this has taken us far from the original issue: What can be said about the overall distribution of primes among the whole numbers? One option is to approach this by collecting data, examining them, and looking for evidence of a possible rule. In this spirit we proceed.

It is customary at this point to introduce the symbol $\pi(x)$ to stand for the number of primes less than or equal to the integer x. For instance, $\pi(8) = 4$ because 2, 3, 5, and 7 are the four primes falling at or below 8. Likewise, $\pi(9) = \pi(10) = 4$ as well. But $\pi(13) = 6$ because 2, 3, 5, 7, 11, and 13 are the six primes less than or equal to 13.

We now gather data. This amounts to counting primes and creating a table for $\pi(x)$. Such a table appears below, where we have listed, among other things, values of $\pi(x)$ for powers of 10 running from 10 to 10 billion.

x	$\pi(x)$	$\pi(x)/x$	$r(x) = x/\pi(x)$
10	4	0.40000000	2.50000000
100	25	0.25000000	4.00000000
1,000	168	0.16800000	5.95238095
10,000	1,229	0.12290000	8.13669650
100,000	9,592	0.09592000	10.4253545
1,000,000	78,498	0.07849800	12.7391781
10,000,000	664,579	0.06645790	15.0471201
100,000,000	5,761,455	0.05761455	17.3567267
1,000,000,000	50,847,534	0.05084753	19.6666387
10,000,000,000	455,052,512	0.04550525	21.9754863
.	.	.	.
.	.	.	.
.	.	.	.

The two rightmost columns of the table require a word of explanation. One gives values of

$$\frac{\pi(x)}{x}$$

which is the *proportion* of numbers less than or equal to x that are prime. For example, that there are exactly 78,498 primes less than or equal to one million so that

$$\frac{\pi(1{,}000{,}000)}{1{,}000{,}000} = \frac{78{,}498}{1{,}000{,}000} = 0.078498$$

This means that 7.85 percent of all numbers below a million are prime; the lion's share, 92.15 percent, are composite.

The far right column gives the reciprocal of

$$\frac{\pi(x)}{x}$$

which we call $r(x)$. For the case of $x = 10$, we see that

$$r(10) = \frac{10}{\pi(10)} = \frac{10}{4} = 2.5$$

The reason for including this column is that we eventually will identify $r(x)$, at least approximately, as a familiar mathematical entity.

What patterns are apparent in the table? It is clear that as x increases the proportion of primes less than or equal to x decreases (scan down the third column). In other words, primes become proportionately rarer as we move into the larger numbers. A moment's reflection suggests the reasonableness of this phenomenon. After all, for a number to be prime it must avoid divisibility by all lesser numbers. For small numbers, having fewer predecessors, such an escape is more likely. Thus for 7 to be prime, it needs only to have no divisors among 2, 3, 4, 5, or 6. But for 551 to be prime, it must avoid divisors among the numbers 2, 3, 4, 5, . . . , 549, and 550, and this seems like a much less likely event. (In fact, 551 is divisible by 19 and so is not a prime.) Just as it is easier to run between raindrops in a light sprinkle than in a raging thunderstorm, so it is easier for a number to be prime if it has fewer, smaller numbers to dodge.

But mathematicians want something stronger than the innocuous observation that the primes get scarcer as we go. They seek a rule or formula that mirrors, at least roughly, the distribution of primes. For this, the table seems to be of little help. Even the most perceptive observer will be forgiven for not spotting a pattern among its entries.

Nonetheless it is there—subtle, sophisticated, and quite unexpected. To spot the pattern, we must again consider the number e and the natural logarithm. It may seem quite fantastic that e has anything to do with the primes, but as we saw in Chapter N this number crops up in the most unexpected places.

So, extend the table to include a column containing the values of $e^{r(x)}$. For instance, when $x = 10$, $r(x) = 10/4 = 2.5$, so we enter the value $e^{2.5} = 12.182494$ in the right-hand column. Proceeding in this fashion yields:

x	$r(x) = x/\pi(x)$	$e^{r(x)}$
10	2.50000000	12.182494
100	4.00000000	54.598150
1,000	5.95238095	384.668125
10,000	8.13669650	3,417.609127
100,000	10.4253545	33,703.4168
1,000,000	12.7391781	340,843.2932
10,000,000	15.0471201	3,426,740.583
100,000,000	17.3567267	34,508,861.36
1,000,000,000	19.6666387	347,626,331.2
10,000,000,000	21.9754863	3,498,101,746.
.	.	.
.	.	.
.	.	.

Although the right-hand column does not display perfect regularity, one senses an underlying principle in action: As we move down, each entry on the right seems to be about ten times as great as the entry above. It appears as though dropping from one line to the next—and thereby increasing the value of x by a factor of 10— roughly increases the value of $e^{r(x)}$ by a factor of 10 as well.

The phenomenon can be summarized by the algebraic expression

$$e^{r(10x)} \approx 10e^{r(x)} \text{ for large } x$$

This simply says that by increasing the input from x to $10x$, the new output, $e^{r(10x)}$, will be about ten times as great as the old output, $e^{r(x)}$.

It may not seem like it, but this observation is significant. We had set as our goal the identification of $r(x)$, and now at least we are in possession of a relevant formula, namely $e^{r(10x)} \approx 10e^{r(x)}$. Surely this is not true of *every* function. If we can find one obeying this rule, we shall have gone a long way toward identifying $r(x)$.

We call upon the natural logarithm. In Chapter N it was stressed that

$$\ln(e^x) = x$$

which says that taking the logarithm undoes the process of exponentiation. But this works in the other direction as well: If we begin with x, take its natural logarithm, and then exponentiate the result, we return to x. Symbolically,

$$e^{\ln x} = x \tag{*}$$

As a numerical example, if $x = 6$, then $\ln x = \ln 6 = 1.791759469$ and $e^{\ln x} = e^{\ln 6} = e^{1.791759469} = 6$. We are back where we started.

So if we begin with $10x$, take the natural log to get $\ln(10x)$, and then exponentiate to get $e^{\ln(10x)}$, the inversion property shows we shall simply get $10x$ again. That is, $e^{\ln(10x)} = 10x$. But by (*) it is clear that $10x = 10e^{\ln x}$. Putting these two facts together, we conclude that

$$e^{\ln(10x)} = 10e^{\ln x}$$

All that remains is to consider this equation alongside the relationship above. That is, we compare

$$e^{r(10x)} \approx 10e^{r(x)} \quad \text{and} \quad e^{\ln(10x)} = 10e^{\ln x}$$

The patterns are identical. We make a bold hypothesis: $r(x)$ is roughly equal to $\ln x$ when x is large.

This is the essence of the prime number theorem, although it is usually recast in a slightly different form. That is, replace $r(x)$ by $x/\pi(x)$ to get $x/\pi(x) \approx \ln x$ and then take reciprocals to end up with:

PRIME NUMBER THEOREM: $\pi(x)/x \approx \dfrac{1}{\ln x}$ for large x. ■

In this form, the theorem is seen in all of its glory. It says that the proportion of primes among the whole numbers, $\pi(x)/x$, is roughly equal to the reciprocal of ln x when x is large. That the distribution of primes should be so linked to the natural logarithm is quite extraordinary.

Of course, we have not *proved* anything. Nor will we. We have simply gotten a sense of what the answer should be. As a kind of numerical check, we modify our table to include both $\pi(x)/x$ and its approximator $1/\ln (x)$:

x	$\pi(x)/x$	$1/\ln (x)$
10	0.40000000	0.43429448
100	0.25000000	0.21714724
1,000	0.16800000	0.14476483
10,000	0.12290000	0.10857362
100,000	0.09592000	0.08685890
1,000,000	0.07849800	0.07238241
10,000,000	0.06645790	0.06204207
100,000,000	0.05761455	0.05428681
1,000,000,000	0.05084753	0.04825494
10,000,000,000	0.04550525	0.04342945
.	.	.
.	.	.
.	.	.

Certainly the agreement is not perfect, but it does seem to improve with increasing x. As the last entry shows, the proportion of primes less than or equal to ten billion differs from $1/\ln (10,000,000,000)$ by only 0.002, so the approximation is off by two parts in a thousand. For some strange reason, as they head off toward infinity, the primes are marching to the beat of the natural log.

If readers believe that no mortal could ever discern such a relationship, they are advised to think again. Among the papers belonging to the 14-year-old Carl Friedrich Gauss, there appeared the following:[1]

$$\text{prime numbers below } a \ (= \infty) \ \frac{a}{la}$$

What is the meaning of these jottings? First, we can replace "prime numbers below a" by its modern equivalent, $\pi(a)$. Moreover, it is obvious that "la" is our "$\ln a$." And surely "$(= \infty)$" means "as $a \to \infty$" or "for large values of a." Thus Gauss's cryptic phrase translates into

$$\pi(a) \approx \frac{a}{\ln a} \text{ for large values of } a$$

We divide both sides by a to arrive at

$$\frac{\pi(a)}{a} \approx \frac{1}{\ln a} \text{ for large values of } a$$

and this is exactly the prime number theorem as stated above! Obviously the adolescent Gauss had recognized the pattern.

It may seem that his achievement was not unlike Houdini's ability to escape from a chained, underwater safe—which is to say, the boy's gifts look like magic. But we must not forget that Gauss had always been fascinated by numbers, that he had an astronomically high I.Q., and that he lived in the days before MTV.

As noted, Gauss recognized the pattern but did not prove it. Nor did anyone else for the next hundred years. At last the prime number theorem was proved by Jacques Hadamard (1865–1963) and C. J. de la Vallee Poussin (1866–1962) in 1896 using some very sophisticated techniques from analytic number theory. Besides sharing almost identical life spans, Hadamard and Vallee Poussin discovered their proofs independently and simultaneously and thus share the honor of establishing this mathematical landmark.

We conclude with a telling observation. Literally thousands of theorems have been proved about prime numbers from the time of Euclid down to the present. Many are significant; some are beautiful. But among them, only one—the object of this chapter—is universally called *the* prime number theorem.

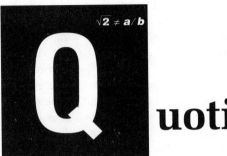

Quotient

In his *Géométrie* of 1637, René Descartes observed, "arithmetic consists of only four or five operations, namely addition, subtraction, multiplication, division, and the extraction of roots."[1]

Apart from his unaccountable waffling on the precise number ("four or five"), Descartes was quite explicit about the allowable operations of arithmetic. From a modern vantage point, these can be used to generate a hierarchy of number systems, each extending its predecessor while simultaneously introducing greater algebraic maneuverability. The construction of number systems from the operations of arithmetic has both logical and historical appeal.

The odyssey begins, as did Chapter A, with the set of natural numbers, denoted by N. Suppose we are working within this system and have at our disposal the lone operation of addition. That is, we are free to choose any two natural numbers, add them, and record the result. If we were to sum all possible pairs and collect the outcomes into a set, what would it be?

The answer is immediate: It would be N again. Mathematicians say the natural numbers are ***closed*** under the operation of addition, by which they mean that one can never escape from N by adding its members. The set N is adequate for adders.

The same can be said if we allow multiplication. Sums and/or products of whole numbers are whole numbers, so N exhibits not only additive but multiplicative closure. So far, so good.

But things deteriorate when we throw in subtraction. It is not true that the difference of two natural numbers is a natural number, for although 2 and 6 are in N, 2–6 is not. With subtraction we can escape.

At this point we face two alternatives. One is suggested by the old joke in which a patient, painfully raising an arm, says "Doctor, it hurts when I do that."

The doctor's advice: "Don't do that."

In this spirit, we could overcome the defects of subtracting natural numbers by forbidding subtraction—"don't do it." This, of course, is ridiculous. The other alternative, and the remedy mathematicians have adopted, is to allow subtraction but enlarge the number system accordingly. The extended system, including negative whole numbers as well as zero, is called the *integers* and is denoted by **Z.**

Although it seems strange to us, many mathematicians were at first violently opposed to such an extension of the idea of "number." This was due in part to the geometrical flavor of mathematics as inherited from the Greeks, for it was difficult to imagine negative lengths, areas, or volumes. And it was due in part to a philosophical aversion to *quantities* being so deficient as to be less than zero. Thus we find Michael Stifel (ca. 1487–1567) calling negative numbers "*numeri absurdi*," and Gerolamo Cardano using the equally scornful term "*numeri ficti*." Opposition continued well into the eighteenth century, as in this passage from Baron Francis Maseres (1731–1824):

> It were to be wished . . . that negative numbers had never been admitted into algebra or were again discarded from it: for if this were done, there is good reason to imagine, the objections which many learned and ingenious men now make to algebraic computations, as being obscure and perplexed with almost unintelligible notions, would be thereby removed.[2]

Even Descartes referred to negatives as *racines fausses*, or false roots. To many mathematicians, there was something strange and disturbing about them.

Nonetheless they cured the ills of subtraction, for any two integers chosen from **Z** and added, subtracted, or multiplied will produce yet another integer in **Z.** The new number system is closed under three of Descartes's operations.

Next comes division, and this presents more fundamental problems. Sometimes things work fine; for instance, 6 and 2 are in **Z,** as is $6 \div 2 = 3$. To use a culinary example, we can divide 6 apples between 2 people by giving each person 3 apples.

But how do we divide 2 apples among 6 people? The answer—as every comedian will attest—is to make applesauce. This is a humorous admission that **Z** is not closed under division, for $2 \div 6$ fails to be in **Z.** (Along these lines, Groucho Marx was once asked how to divide 2 *umbrellas* among 6 people. His answer: Make umbrella-sauce.)

To accommodate division is no joking matter. It requires another extension of the number system to the set of quotients or, in more technical jargon, ***rational numbers.*** Formally, the rationals **Q** is the set of all quotients a/b where a and b are integers and $b \neq 0$. Thus $-2/3$ is in **Q,** as are $7/18$ and $18/7$. Note that any integer a also lives in **Q** because $a = a/1$, where the latter is in fractional form.

The one restriction above is that the denominator of a fraction cannot be zero. No expression such as $4/0$ is allowed among the rationals. To see why not, suppose for a moment that $4/0$ *does* have meaning so there exists a number x with $4/0 = x$. If we

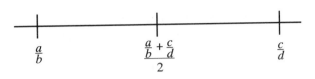

Figure Q.1

cross-multiply, we get $4 = 0 \times x$. But, of course, $0 \times x = 0$, so we are left with $4 = 0$, an unacceptable state of affairs by anyone's standards. Mathematicians conclude that a quotient with zero in the denominator is not a quotient at all. Dividing by zero is the closest thing there is to arithmetic blasphemy.

Two important properties of the rational numbers deserve mention. They are significant because they are not shared by the natural numbers or the integers and illustrate a superiority of **Q** over its predecessors **N** and **Z**.

The first is that **Q** is closed under the four basic operations of addition, subtraction, multiplication, and division, except of course the forbidden division by zero. Mathematicians favor number systems of this sort, for they can add, subtract, multiply, and divide to their heart's content while remaining within the system.

The second major distinction of the rationals is that they are densely distributed. This means that between any two fractions must lie another. The integers clearly do not possess this property because the gap between, for instance, 5 and 6, is integer-free. Integers are arranged in lockstep fashion, each having its successor a full unit away. We think of them as scattered, isolated, and discrete.

Not so the quotients. Between 1/2 and 4/7 lies 15/28, between 15/28 and 4/7 is 31/56, and so on. In general, for any two rationals

$$\frac{a}{b} < \frac{c}{d}$$

their average value

$$\frac{\frac{a}{b} + \frac{c}{d}}{2}$$

falls halfway between them (see Figure Q.1). Moreover, by multiplying numerator and denominator of this average by bd, we see that

$$\frac{\frac{a}{b} + \frac{c}{d}}{2} = \frac{\frac{a}{b} + \frac{c}{d}}{2} \times \frac{bd}{bd} = \frac{ad + bc}{2bd}$$

so that the average of two rationals is indeed another rational.

Because this process can be repeated indefinitely, it is clear that between any two rationals lies an *infinitude* of rationals. In this sense rational numbers are more densely packed than any can of sardines or jar of pickles. They are abundant beyond comprehension.

Does this mean that *all* numbers are rational? The answer is "no," although it is not absolutely trivial to see this. One route is to consider the representation of fractions as infinite decimals.

We all remember decimal expansions from elementary school. The usual pencil-and-paper calculations generate a chain of quotients and remainders, as illustrated in the determination of the decimal for 5/8:

$$
\begin{array}{r}
0.625 \\
8\overline{)5.000} \\
\underline{48} \\
20 \\
\underline{16} \\
40 \\
\underline{40} \\
0
\end{array}
$$

Here we have highlighted the remainders as we moved downward: 5 (where we began), then 2, 4, and 0. Once a zero remainder appears, there is nothing more to be done, and we end up with 5/8 = 0.625. Because we are trying to represent rationals as *infinite* decimals, we can attach an infinite string of zeros and write 5/8 = 0.625000 . . .

This example, in which the division process stops, illustrates one of two alternatives. The other arises when we compute the decimal for a rational number such as 5/7:

$$
\begin{array}{r}
0.714285 \ldots \\
7\overline{)5.0000000} \\
\underline{49} \\
10 \\
\underline{7} \\
30 \\
\underline{28} \\
20 \\
\underline{14} \\
60 \\
\underline{56} \\
40 \\
\underline{35} \\
50
\end{array}
$$

This time the process shows no sign of stopping. But as we consider the string of remainders—5, 1, 3, 2, 6, 4, 5—we see that a repetition occurs. At this point we will again find ourselves dividing 7 into 50, so we must run through the same cycle. The decimal expansion will then repeat the digits 714285, only to return to a remainder of 5 and start another cycle. The expansion of 5/7 is

$$0.714285714285714285714285\ldots$$

The critical question is whether the repetition was a lucky break or a general principle. It is easy to see that it is the latter. When dividing by 7, the only possible remainders are 0, 1, 2, 3, 4, 5, or 6. If the remainder is ever zero, the process stops. Otherwise, after no more than six steps we *must* get a remainder we have previously seen, for the digits from 1 to 6 can only be spread around so far. Once a remainder repeats, so will the cycle of divisors.

There is nothing special about the divisor 7. Precisely the same reasoning shows that when converting 113/757 to a decimal, the expansion must repeat after at most 756 steps (it actually repeats much sooner). Generally, when attacking a/b, the division either stops or repeats after at most $b - 1$ steps.

We thus see that the decimal expansion of any rational number *must* feature a repeating block. This is true whether the repeating block is "0" as in the first example or "714285" as in the second. Rationals are repeaters.

This criterion provides a recipe for generating a number that is not rational: Simply create an infinite decimal *without* a repeating block. For instance, the real number

$$0.10100100010000100000100000010000000\ldots$$

has strings of one 0, then two 0s, then three, and so on. No block repeats. This number, therefore, is not rational and so *cannot* be expressed as the quotient of two integers. It is what mathematicians call an **irrational number** and illustrates what Stifel meant by his perceptive observation "when we seek to subject them [irrationals] to numeration . . . we find that they flee away perpetually, so that not one of them can be apprehended precisely in itself." The endless irregularity in the decimal expansion of irrationals shows that, in Stifel's words, irrationals are hidden "in a cloud of infinity."[3]

Although the decimal cited above is an irrational, it is not one that held any prior interest. Far more intriguing would be the identification of a prominent, widely used number as an irrational. Such a number is $\sqrt{2}$, whose irrationality was recognized 25 centuries ago by the Pythagoreans of ancient Greece.

The importance of $\sqrt{2}$ is evident if we consider a square of side s and diagonal d in Figure Q.2. The Pythagorean theorem states that

$$d^2 = s^2 + s^2 = 2s^2 \quad \text{and so} \quad d = \sqrt{2s^2} = s\sqrt{2}$$

Thus $\sqrt{2}$ is present in any square, be it a highway sign, a checkerboard, or a baseball diamond.

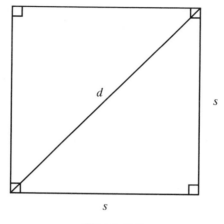

Figure Q.2

In spite of its prominence, $\sqrt{2}$ is an irrational number. This fact seems to have come as an unwelcome surprise to the Greeks, and legend says that its discoverer, Hippasus the Pythagorean, was murdered by his mathematical companions for publicizing something so unfortunate. A result with life or death consequences deserves special attention, so we shall give two different irrationality proofs. One requires a bit of geometry and the other a bit of number theory. In both, our objective is to show that $\sqrt{2}$ cannot be expressed as the quotient of two integers, no matter how diligently we try.

As we noted in Chapter J, an objective of this sort can never be achieved by checking a few specific examples. A chemist who drops 50,000 pieces of sodium into 50,000 beakers of water and witnesses 50,000 explosions may properly conclude that something has been established. But a mathematician who checks 50,000 fractions and finds that none of them equals $\sqrt{2}$ is nowhere nearer a general result than if he or she had never begun.

For the matter at hand, a more sophisticated weapon is necessary. That weapon is proof by contradiction, which will undergird the two arguments that follow. In both cases, to prove that $\sqrt{2}$ is irrational, we begin by assuming the opposite, that $\sqrt{2}$ is a rational number, and from this derive a contradiction.

THEOREM: $\sqrt{2}$ is irrational.

PROOF: (by contradiction) Suppose that $\sqrt{2}$ is rational. Then there exist positive integers a and b with $\sqrt{2} = a/b$. Here we require—and this is crucial—that a/b be reduced to lowest terms. This is not an unreasonable demand, for it is always possible to so adjust a fraction. (For instance, we can reduce 15/9 to 5/3.)

With $\sqrt{2} = a/b$ in lowest terms, construct a square b units on a side (see Figure Q.3). By our earlier observation, the diagonal has length $b\sqrt{2} = b \times (a/b) = a$. Along

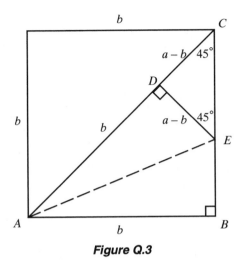

Figure Q.3

this diagonal mark off segment AD of length b and draw $DE \perp AC$, where E is on BC as shown. Then segment CD has length $\overline{AC} - \overline{AD} = a - b$.

Note that $\angle ACB$ contains 45° and $\angle CDE$ contains 90°, so the remaining angle of $\triangle CED$ also measures 45°. This means $\triangle CED$ is isosceles, and it follows that $\overline{ED} = \overline{CD} = a - b$.

Next draw segment AE, forming right triangles ADE and ABE with shared hypotenuse AE. Applying the Pythagorean theorem to both yields

$$\overline{AB}^2 + \overline{EB}^2 = \overline{AE}^2 = \overline{AD}^2 + \overline{ED}^2$$

Thus $b^2 + \overline{EB}^2 = b^2 + \overline{ED}^2$, from which it follows that $\overline{EB}^2 = \overline{ED}^2$ and so $\overline{EB} = \overline{ED} = a - b$. Consequently, $\overline{EC} = \overline{BC} - \overline{EB} = b - (a - b) = 2b - a$.

We now focus on the small right triangle CED. Because its two legs are of length $a - b$, the Pythagorean theorem implies that its hypotenuse EC is of length $(a - b)\sqrt{2}$. On the other hand, we have just shown that EC has length $2b - a$. Hence $(a - b)\sqrt{2} = 2b - a$, or simply

$$\sqrt{2} = \frac{2b - a}{a - b}$$

To summarize the argument thus far, we assumed $\sqrt{2} = a/b$ in lowest terms and then, using some elementary geometry, deduced that

$$\sqrt{2} = \frac{2b - a}{a - b}$$

as well. It may not seem like it, but we are on the brink of a contradiction. To wrap this up we need four simple observations:

1. Because a and b are integers, so, too, are $2b - a$ and $a - b$.

2. Because we assumed $\sqrt{2} = a/b$, and because it is evident that $1 < \sqrt{2} < 2$, we know that $1 < a/b < 2$. Multiplying through by b yields the inequality $b < a < 2b$.

3. From the fact that $b < a$ in item (2), we deduce that $a - b$ is positive, and from the fact that $a < 2b$ in item (2), we deduce that $2b - a$ is also positive.

4. Doubling both sides of the inequality $b < a$ in item (2), we see that $2b < 2a$, and upon subtracting a from both sides, we arrive at $2b - a < a$.

So, although we had assumed that $\sqrt{2} = a/b$ is written as a quotient of positive integers *in lowest terms*, we now have

$$\sqrt{2} = \frac{2b - a}{a - b}$$

where both numerator and denominator are positive integers (by item (1) and item (3)), but where the new numerator $2b - a$ is strictly less than the original numerator a (by item (4)). Because we now have a smaller numerator, we have reduced the fractional representation for $\sqrt{2}$ beyond what was assumed to have been its lowest terms, a reduction that is obviously impossible.

The contradiction has arrived (and none too soon). It follows that this entire argument was built upon the quicksand of its initial erroneous assumption, that $\sqrt{2}$ was rational. Rejecting this, we conclude that $\sqrt{2}$ is an irrational, and the proof is complete. ■

At the risk of beating a dead number, we now give a second and much shorter proof of the irrationality of $\sqrt{2}$. Its prerequisites are the unique factorization theorem for whole numbers, discussed in Chapter A, and one additional principle we now develop.

Suppose a positive integer m has been factored into primes. For instance, if $m = 360$, we have $m = 2^3 \times 3^2 \times 5$. Note that the prime 2 appears three times in this factorization, the prime 3 appears twice, the prime 5 appears once, and all other primes—7, 11, 13, etc.—make no appearance whatever. Certainly for an unspecified integer there is not much we can say about the number of occurrences of any given prime.

But consider m^2. For our example, this would be

$$m^2 = 360^2 = (2^3 \times 3^2 \times 5) \times (2^3 \times 3^2 \times 5) = 2^6 \times 3^4 \times 5^2$$

Note that in the prime factorization of m^2, each prime occurs twice as often as it did in the factorization of m, which means that each prime occurs an *even* number of times. Thus, the three 2s from the first m multiply the three 2s from the second m to yield six 2s in m^2. Similarly, there are four 3s, two 5s, and no 7s, 11s, 13s, and so on.

A little thought reveals that this phenomenon will always occur, because by squaring any positive integer we double the number of appearances of its primes and doubling leads to an even result. (Zero, being the double of itself, is also even.) In short, we have verified the following principle:

> The square of any whole number must have each prime occurring an *even* number of times in its prime factorization.

With this established, we are ready for our second irrationality proof. ⸰

THEOREM: $\sqrt{2}$ is irrational.

PROOF: (by contradiction) Suppose that $\sqrt{2}$ is rational. Then there exist whole numbers a and b for which $\sqrt{2} = a/b$. Square both sides and cross-multiply to get $2b^2 = a^2$.

We consider the prime factorization of both sides of this equation. On the right appears a^2, the square of integer a. By the principle above, we know the prime 2 must occur an even number of times in this factorization. Meanwhile the left-hand side contains b^2 and again the prime 2 must appear an even number of times in its factorization. But—and here's the rub—the *entire* left side is $2b^2$, so an additional 2 is present. Because b^2 contributes an even number of 2s, the expression $2b^2$, when factored into primes, must contain an *odd* number of 2s.

We have thus reached a contradiction, for we have shown that a^2, when factored into primes, contains an even number of 2s, yet the prime factorization of the same number $2b^2$ contains an odd number of 2s. Thus our number—call it a^2 or $2b^2$—has two *different* prime factorizations.

This is impossible, for it contradicts the unique factorization theorem from Chapter A. Logically, something has gone awry. A look back through the chain of reasoning shows that trouble entered with the initial assumption that $\sqrt{2}$ can be written as a fraction. We jettison this assumption and conclude again that $\sqrt{2}$ is irrational. ■

This second proof, with 2 replaced by 3, 5, 7, or any other prime, demonstrates the irrationality of $\sqrt{3}$, $\sqrt{5}$, $\sqrt{7}$, and so on. It turns out that for any whole number n that is not a perfect square, $\sqrt{n}$ is irrational. So too are numbers like $\sqrt[3]{n}$ where n is not a perfect cube, $\sqrt[4]{n}$ where n is not a perfect fourth power, and so on.

So, irrationals abound. But these examples of irrationals, although not expressible as quotients of integers, are "tame" by at least one standard: All are solutions of simple polynomial equations with integer coefficients. For instance, the irrational $\sqrt{2}$ is the solution of the quadratic $x^2 - 2 = 0$, and the irrational $\sqrt[4]{7}$ solves the fourth-degree equation $x^4 - 7 = 0$. Both equations, it must be stressed, have integers as coefficients.

A number of this sort is called **algebraic**. It follows that any rational number a/b is algebraic, for it solves $bx - a = 0$, a first-degree equation whose coefficients b and $-a$ are integers. In this sense the algebraic numbers can be regarded as an extension of the rationals that arise simply by removing the "first-degree" restriction and allowing polynomials of any degree whatever, provided their coefficients remain integers.

Although any rational number is algebraic, it is certainly not true that any algebraic number is rational, as $\sqrt{2}$ or $\sqrt[4]{7}$ illustrate. The algebraic numbers contain all of the rationals and a multitude of irrationals as well. For instance, we assert that the irrational $1/2 + \sqrt{11}$ is algebraic. To verify this, we must produce a specific polynomial equation for which it is a solution. This is done as follows: Start with $x = 1/2 + \sqrt{11}$ and work backwards algebraically to eliminate the radical. That is,

$$x - \frac{1}{2} = \sqrt{11} \rightarrow \left(x - \frac{1}{2}\right)^2 = (\sqrt{11})^2 = 11$$

Expanding the left side gives $x^2 - x + 1/4 = 11$. Then, to meet the requirement that all coefficients be *integers*, multiply both sides by 4 and collect terms to get

$$4x^2 - 4x - 43 = 0$$

A quadratic polynomial equation with integer coefficients has been constructed with $1/2 + \sqrt{11}$ as a solution. By definition, this means that $1/2 + \sqrt{11}$ is algebraic.

A similar strategy shows that a complicated expression such as

$$\frac{\sqrt{6}}{\sqrt[3]{5} + \sqrt{3}}$$

is algebraic, for it is the solution of

$$4x^{12} - 49{,}248x^{10} - 37{,}260x^8 - 127{,}440x^6 + 174{,}960x^4 - 139{,}968x^2 + 46{,}656 = 0$$

although such a derivation is not for the timid.

Any real number created in a finite number of steps from the integers and Descartes's five allowable operations of addition, subtraction, multiplication, division, and root extraction will be algebraic. It is, frankly, difficult to imagine a nonalgebraic number, one that is not the solution of *any* polynomial equation with integer coefficients.

It was Euler who first speculated that such numbers exist. A real number that is not algebraic he called **transcendental** because it transcends the operations of algebra.[4] Transcendentals were thus introduced not by saying what they are but by saying what they are not: They are not algebraic.

Defining transcendentals in this negative fashion leaves open the question of existence. For instance, we could analogously define a dolphin to be *algebraic* if it lives in water and *transcendental* if it is not algebraic. Logically, this definition is fine; but, of course, there are no transcendental dolphins.

Are there transcendental *numbers*? Euler could not find any. It took a century until Joseph Liouville (1809–1882) concocted a number whose transcendence he could prove. Defined by an infinite series, his example was:

$$\frac{1}{10^{1!}} + \frac{1}{10^{2!}} + \frac{1}{10^{3!}} + \frac{1}{10^{4!}} + \frac{1}{10^{5!}} + \ldots = \frac{1}{10} + \frac{1}{10^2} + \frac{1}{10^6} + \frac{1}{10^{24}} + \frac{1}{10^{120}} + \ldots$$

$$= 0.1 + 0.01 + 0.000001 + 0.0000000000000000000000001 + \ldots$$

$$= 0.110001000000000000000001000000 \ldots$$

where the 1s become increasingly rare as we move rightward. Liouville's proof was a brilliant piece of work, and it established once and for all that transcendental numbers exist. Yet it was difficult not to feel a bit crestfallen that the first known transcendental number turned out to be one that was so artificial.

It would be far more satisfying to prove the transcendence of a famous, well-established number, and two candidates soon caught the fancy of mathematicians: the circular constant π that we met in Chapter C and the natural growth constant e that appeared in Chapter N.

It had long been known that both π and e were irrational. The irrationality of e had been recognized by Euler himself as early as 1737, and the irrationality of π was established by Johann Lambert (1728–1777) in 1767. But being irrational does not mean a number is transcendental (think of the irrational but algebraic number $\sqrt{2}$). Proving transcendence is a vastly more difficult undertaking.

It was e that fell first. After intense effort, Charles Hermite (1822–1901) proved that e was transcendental in 1873. His result was properly hailed as a triumph of mathematical reasoning. Unlike Liouville, who *created* a number whose transcendence he could prove, Hermite had to wrestle with an already specified opponent. Liouville was like a paleontologist asked to find a dinosaur bone but permitted to search the world over; Hermite, by contrast, was told to find a Tyrannosaurus skull in his backyard.

But he did it. On the heels of his triumph Hermite was urged to tackle π. He declined, and his words suggest the intellectual toll these efforts had exacted: "I do not dare to attempt to show the transcendence of π. If others undertake it, no one will be happier than I about their success, but believe me, my dear friend, this cannot fail to cost them some effort."[5] Hermite, great mathematician though he was, wanted no part of proving the transcendence of any more numbers. One such ordeal was enough.

And so it fell to Ferdinand Lindemann (1852–1939) to polish off the transcendence of π in 1882. Ironically, Lindemann's proof rested upon Hermite's groundbreaking—and mindbreaking—work and turned out to be easier than expected.

With this, the two great constants e and π were shown to be not merely irrational but far worse: Neither was the solution of any polynomial equation with integer coefficients. If Stifel correctly characterized irrationals as being hidden under a "cloud of infinity," the transcendentals seem to be hidden under a cloud of algebraic inaccessibility.

So where does this leave us? Descartes's short list of algebraic operations opened the door to various classes of numbers, from the simple integers to the quotients mentioned in this chapter's title. But we have seen that quotients are not sufficient to embrace the irrationality of $\sqrt{2}$, and Hermite and Lindemann proved that no amount of adding, subtracting, multiplying, dividing, or root taking would yield numbers such as e and π. The discovery of transcendentals, like the earlier discovery of irrationals, showed that the real numbers are a much stranger and more complicated system than anyone at first imagined.

ussell's Paradox

Bertrand Arthur William Russell was born at the customary age of 0 years on the eighteenth of May, 1872, and died at the atypical age of 97 years on the second of February, 1970. For nearly a century he lived a life of astonishing richness and turbulence, achieving fame as a philosopher and social critic, as a writer and educator, as a member of the House of Lords and an inmate of Brixton Prison. He taught at many of the world's most prestigious institutions, from Cambridge to Harvard to Berkeley. He won a Nobel Prize. He also had four marriages and a number of affairs. He was reviled for his godless agnosticism and his advocacy of sex outside the bonds of marriage. A list of those he encountered along the way reads like a who's who of Western civilization.

The first part of this chapter sketches Bertrand Russell's extraordinary life. In so doing, we draw heavily from his own writings and from Ronald Clark's excellent 1976 biography, *The Life of Bertrand Russell*. Then we return to examine Russell's paradox, one of his early discoveries that had an unsettling impact upon the foundations of mathematics early in the twentieth century. In this way, we hope to give a sense of the man and his work.

One of the striking things about Russell was his strange mixture of conformity and nonconformity, of traditional values and shocking radicalism. In some ways he seemed very much a product of the British upper crust; in other ways he seemed an eternal enemy of the status quo. Photographs show him leading antiwar protests in a three-piece suit and watch fob. Although his vow not to "respect respectable people" must have labeled him a traitor to his class, Bertrand Russell had a background as respectable as they come.[1]

His grandfather John Russell had been Queen Victoria's prime minister from 1846 to 1852 and again from 1865 to 1866. Bertrand, who would live to see humans walking on the moon, recalled sitting upon the royal knee of Victoria when she visited his grandfather's estate. Clearly young Bertie was born into the highest echelons of nineteenth-century British society.

Yet even for the powerful, life can be cruel. Russell lost both parents by the age of four. As a consequence he was raised primarily by his grandmother, who decided to educate him not at school but by tutors at home. This bright and sensitive lad thus spent much of his youth living among elders in the silent ancestral mansion at Pembroke Lodge, deprived of the carefree joys of childhood. By his own account, he was a lonely and repressed youth who spent an inordinate amount of time brooding. He brooded about good and evil and more than once contemplated suicide.

But from this solitary childhood Russell took one lesson that he would carry with him to the end. It was his grandmother's favorite biblical passage—"Thou shalt not follow a multitude to do evil"—words that would serve to characterize Russell's life about as well as any.[2]

When the time came, Bertie left Pembroke Lodge for Trinity College, Cambridge, the same institution that had welcomed a young Isaac Newton more than two centuries before. With his restricted background and intellectual intensity, he came across as a rather odd bird. But the scholarly life suited him well, and mathematics caught his interest above all.

It was love at first sight. Russell felt himself woefully inadequate in the physical or experimental sciences, but mathematics—an impersonal subject that, in his words, he could love but that would not love him in return—became an obsession. For Russell, mathematics offered the one route to certainty and perfection. "I disliked the real world," he confessed, "and sought refuge in a timeless world, without change or decay or the will-o'-the-wisp of progress."[3] In this spirit, he wrote the following paean to mathematics, a tribute whose excess is tempered only by its eloquence:

> Real life is, to most men, a long second-best, a perpetual compromise between the ideal and the possible; but the world of pure reason knows no compromise, no practical limitations, no barrier to the creative activity embodying in splendid edifices the passionate aspiration after the perfect from which all great work springs. Remote from human passions, remote even from the pitiful facts of nature, the generations have gradually created an ordered cosmos, where pure thought can dwell as in its natural home, and where one, at least, of our nobler impulses can escape from the dreary exile of the actual world.[4]

As might be surmised from these words, the utilitarian aspects of mathematics held little attraction. His love was for a purer, more ascetic kind of mathematical reasoning. In his *Introduction to Mathematical Philosophy*, Russell described the two great and opposite directions of mathematical thought: "the more familiar . . . is constructive, towards gradually increasing complexity: from integers to fractions, real numbers, complex numbers; from addition and multiplication to differentiation and integration and on to higher mathematics. The other direction, which is less familiar, proceeds . . . to greater and greater abstractness and logical simplicity."[5] It was this other direction, the movement away from applications and complexity and towards foundations and simplicity, that characterized for Russell mathematical philosophy. It was here he found his intellectual home.

His work on the foundations of mathematics was done at Cambridge, first as a student and then as a Fellow. In this undertaking he was joined by Alfred North Whitehead, an established logician whose collaboration with Russell would span decades of scholarly and personal strife. During the summer of 1900, a time of "intellectual intoxication," Russell made significant advances in mathematical logic. It was a heady and exciting period for the 28-year-old intellectual, who later recalled, "I went about saying to myself that now at last I had done something worth doing, and I had the feeling that I must be careful not to be run over in the street before I had written it down."[6]

In 1903, Russell published a 500-page work, *The Principles of Mathematics*, and later he and Whitehead authored the massive *Principia Mathematica* in three volumes appearing in 1910, 1912, and 1913. This was their definitive attempt to reduce all of mathematics to the basic and irrefutable ideas of logic. The *Principia* was so filled with logical symbols to the exclusion of English words that the mathematics historian Ivor Grattan-Guinness aptly described a typical page as looking like "wallpaper."[7] (An excerpt from the work appears in Chapter J.)

The unforgiving exactitude of these volumes drained the reserves of Russell, Whitehead, and presumably of anyone with the perseverance to read them. It also drained their pocketbooks, for very few people wished to purchase so horrific a publication. "We thus earned minus £50 each by ten years' work," Russell conceded.[8] Worse, it was not clear that Russell and Whitehead had succeeded in their mission of reducing all of mathematics to logic. What *was* clear was that they had produced a work probing the foundations of mathematics to unparalleled depths.

On the eve of World War I the 40-year-old Bertrand Russell had thus made a mark in mathematical philosophy. A contemporary might have guessed his remaining years would be spent exploring further arcane theorems of logic. But the contemporary would have guessed wrong, for Russell's life was about to move in remarkable and unexpected directions.

Many forces, both internal and external, propelled him, but foremost among these was the insanity of the First World War. Russell, like many British intellectuals, watched as an entire generation of young men was swept away in the carnage. Suddenly the march of logical symbols across a page lost its importance. He confessed that, in the face of war, "The work I have done seems so little—so irrelevant to this world in which we find we are living."[9]

Bertrand Russell plunged into the fray. His activism against the war led to his arrest in 1916, his dismissal from Cambridge, and the loss of his passport. The last cost him a position that awaited at Harvard. But none of this silenced his scathing denunciation of a war effort that daily grew more tragic, and it was inevitable that a further clash lay in store. This came in 1918 when Russell was again arrested and carted off for six months to Brixton Prison. The son of nobility had become a prisoner of conscience.

It was not his antiwar stance alone that put him at odds with the British establishment. There were at least two other positions that antagonized traditional values. One was his public agnosticism. Russell was a critic not only of specific religions but of religion generally. He was a man who believed above all in the supremacy of reason and saw theology as leading humanity in opposite and unfortunate directions. His denunciations were cutting, powerful, and harsh. He wrote, for instance, that "the more intense has been the religion of any period and the more profound has been the dogmatic belief, the greater has been the cruelty."[10] He regularly attacked the Roman Catholic Church for its prohibition against birth control, and he was barely kinder to the other denominations of Christendom. To those who saw God's handiwork in the design of our universe, Russell asked, "Do you think that, if you were granted omnipotence and omniscience and millions of years in which to perfect your world, you could produce nothing better than the Ku Klux Klan or the Fascists?"[11] His views may be summarized in his answer to the question of what it was in this world that he particularly liked: "Mathematics and the sea, and theology and heraldry, the two former because they are inhuman, the two latter because they are absurd."[12] It is perhaps understandable that, when he was incorrectly reported to have died on a trip to China, a religious journal uncharitably editorialized, "Missionaries may be pardoned for breathing a sigh of relief at the news of Mr. Bertrand Russell's death."[13]

But if his religious views were controversial, so, too, were his views on sex and marriage. There was little in his straightlaced upbringing to predict such unorthodoxy. At the age of 22 he married Alys Pearsall Smith, an American Quaker living in England. Alys insisted upon a Quaker wedding, to which Bertie agreed with characteristic tact; "Don't imagine that I really seriously mind a religious ceremony ... *any* ceremony is disgusting."[14]

At the outset their marriage promised to be eternal, but in matters of the heart there was little of permanence for Bertrand Russell. One day early in 1902, while on a bicycle ride near Cambridge, Russell suddenly realized that he no longer loved his wife.

With that realization began a series of romantic entanglements that would cover half a century and involve this man of reason in behavior that seemed to all the world as decidedly unreasonable. He apparently became infatuated with Evelyn Whitehead, the wife of the man with whom he was writing *Principia Mathematica*. He had a long-running affair with Lady Ottoline Morrell, a well-known British socialite and wife of a prominent politician. There were any number of clandestine meetings in obscure hotel rooms. It was all quite unseemly for a man of international stature.

While all of this was going on, he divorced Alys and married Dora Black in 1921. On paper their marriage lasted until 1935, but in 1929 Russell was writing of his second wife, "Neither she nor I make any pretense of conjugal fidelity."[15] Under these circumstances, he could hardly have been surprised when Dora had another man's baby in 1930. But when she had a *second* child by the same man, even Russell had had enough. He sued for divorce.

This paved the way for his third marriage, to Helen Patricia Spence, which lasted from 1936 until 1952. Then, at the age of 80, he wed Edith Finch, a professor of English at Bryn Mawr, and thus found a partner with whom he could happily spend his final years.

Such marital and extramarital behavior got Bertrand Russell in a great deal of hot water, especially because he was ever ready to discuss his views on sex, chastity, contraception, and the like. In a much celebrated 1940 case, he was banned from a faculty position at City College of New York at the behest of the religious community and Mayor Fiorello LaGuardia. It was announced that Russell was unfit to teach because of his views opposing religion 'and endorsing promiscuity. As a defense of sorts, he once observed that lovestruck mathematicians werê no different than anyone else "except, perhaps, that the holiday from reason makes them passionate to excess."[16] Bertrand Russell obviously spent considerable time on holiday.

But he also spent considerable time at work. During these years of controversy he remained a prolific writer, producing volumes of social commentary, treatises on education, and even articles for the popular press. It seems a bit incongruous, yet this social activist occasionally found himself writing for *Glamour* magazine and appearing as a celebrity guest on a BBC radio program. Part of his popular acceptance was due to the fact that, his views notwithstanding, Bertrand Russell was a genuinely fascinating individual. Part was undoubtedly due to the fact that he outlived his enemies.

Two other aspects of his life deserve mention. One was his abiding distaste for the communist political system. During a time when many intellectuals cheered the rise of communism as the salvation of humanity, Russell characteristically swam against the tide. On purely intellectual grounds, he gave two succinct reasons for opposing the philosophy of Karl Marx: "one, that he was muddle-headed; and the other, that his thinking was almost entirely inspired by hatred."[17]

Russell's disdain for communism went right to its source, for he had met Lenin personally during a 1920 visit to Moscow and come away shaken. His assessment was as harsh as that of any hawkish Western politician when he described the Soviet state as "an asylum of homicidal lunatics where the warders are the worst."[18] During World War II, an effort Russell supported, he wondered whether England's enemy Hitler was really much worse than their ally Stalin.

The other striking feature of Bertrand Russell was his gift as a writer. As noted, he wrote on a wide variety of subjects. But whether philosophical tomes (e.g., *Our Knowledge of the External World as a Field for Scientific Method in Philosophy*), critical tracts (e.g., *An Outline of Intellectual Rubbish*), or popular fluff (e.g., *If You Fall in Love with a Married Man*), the writing was fresh, provocative, and engaging.

And it had an undeniable flair, particularly if tinged with a touch of his biting sarcasm. When writing about the classification of gluttony as a sin, Russell mused: "it is a somewhat vague sin, since it is hard to say where legitimate interest in food ceases and guilt begins to be incurred. Is it wicked to eat anything that is not nourishing? If so, with every salted almond we risk damnation."[19] He poked fun at staunch supporters of animal rights when he wrote: "A resolute egalitarian . . . will find himself forced to regard apes as the equals of human beings. And why stop with apes? I do not see how he is to resist an argument in favour of Votes for Oysters."[20] And he once delayed the writing of an autobiography because: "I have a certain hesitation in starting . . . too soon for fear of something important having not yet happened. Suppose I should end my days as President of Mexico; the biography would seem incomplete if it did not mention this fact."[21]

His talent with the written word was recognized in the most public fashion imaginable when Bertrand Russell received the 1950 Nobel Prize for literature. But, in describing his formula for successful writing, Russell provided little comfort for composition teachers:

> He [a mentor] gave me various simple rules, of which I remember only two: "Put a comma every four words," and "never use 'and' except at the beginning of a sentence." His most emphatic advice was that one must always re-write. I conscientiously tried this, but found that my first draft was almost always better than my second. This discovery has saved me an immense amount of time.[22]

Throughout his life, from his mathematical research to his imprisonment, from his numerous affairs to his Nobel Prize, Russell hobnobbed with a remarkable array of interesting and influential people. His godfather was John Stuart Mill. We already noted that he once sat on Queen Victoria's knee. He later enjoyed the company of John Maynard Keynes, William James, and H. G. Wells. He knew writers Beatrix Potter, D. H. Lawrence, George Bernard Shaw, Joseph Conrad, Aldous Huxley, and Rabindranath Tagore. Among his students were Ludwig Wittgenstein and T. S. Eliot. In Russia he interviewed Lenin and Trotsky, and his 1920 lectures in Beijing were reportedly attended by two intense young radicals, Mao Tse-tung and Cho En-lai. He was friends with everyone from Albert Einstein to Peter Sellers to Winston Churchill. In regard to the last, Russell reported that one evening at a dinner party, "Winston asked me to explain differential calculus in two words, which I did to his satisfaction."[23]

As if this were not adequate contact with greatness, Bertrand Russell occupied the rooms at Trinity College that had once been the residence of Isaac Newton. Although temperamentally Newton and Russell could not have been less alike, these two Englishmen each possessed an intellect of enormous power and each pushed the mathematics of their day to new frontiers.

It is one of these frontiers that we wish to examine. We look back to 1901, when Russell was deep in his research on the logical foundations of mathematics. This necessitated that he examine relationships among collections of things (Russell talked of *classes* although the modern term is *sets*). The nature of the "things" in the classes was immaterial; what mattered was the abstract logic of set theory.

Set membership seems a trivial matter. If we consider the set $\mathbf{S} = \{a, b, c\}$, then b is a member of $\mathbf{S}$ but g is not. If we consider the set of all even whole numbers, then 2, 6, and 1,660 are members of the set, whereas 3, $1/2$, and π are not.

Pushing the level of abstraction a bit, we observe that the members of a set may themselves be sets. For the two-member set $\mathbf{T} = \{a, \{b,c\}\}$, the first member is a and the second member is the *set* $\{b, c\}$. Or suppose we let $\mathbf{W}$ be the set consisting of the set of all even whole numbers and the set of all odd whole numbers. That is,

$$\mathbf{W} = \{\{2, 4, 6, 8, \ldots\}, \{1, 3, 5, 7, \ldots\}\}$$

The set $\mathbf{W}$ has two members, each of which is itself a set containing an infinitude of members.

The fact that a set may have sets as members raised for Russell an intriguing question: Can a set have *itself* as a member? He wrote, "it seemed to me that a class sometimes is, and sometimes is not, a member of itself."[24]

He gave as an example the set of all teaspoons, which is certainly not a teaspoon. Thus, the set of all teaspoons is not a member of itself. Likewise, the set of all people is not a person and hence is not a member of itself.

On the other hand, it seemed to Russell that certain sets *do* contain themselves as members. His example was the set of all things that are not teaspoons. This set of nonteaspoons contains forks, British prime ministers, eight-digit numbers—indeed, anything that is not a teaspoon. But the set itself is surely not a teaspoon (one could not stir tea with it) and so it rightly belongs within itself as yet another nonteaspoon.

Or consider the set $\mathbf{X}$ of all sets that can be described in 20 or fewer English words. The set of all buffaloes would be a member of $\mathbf{X}$ because its description, "The set of all buffaloes," requires only five words. Likewise, the set of all porcupine needles (six words) is in $\mathbf{X}$, as is the set of all mosquitoes living in South America (nine words). But this membership criterion guarantees that $\mathbf{X}$—the set of all sets that can be described in 20 or fewer English words—having thereby been described in 15 words, must be included within itself.

Clearly every set falls into one of two categories. Either it is a set, like the teaspoons, that is not a member of itself—in which case we shall call it a ***Russell set***—or it is a set, like $\mathbf{X}$, that does contain itself among its members.

These innocent reflections took an ominous turn when Russell decided to consider the set of *all* those sets that are *not members of themselves*. That is, collect all of the Russell sets into a huge new set we shall denote as $\mathbf{R}$. Then $\mathbf{R}$ will have among its members the set of all teaspoons, the set of all people, and much, much else.

And now comes the question that shook the foundations: Is $\mathbf{R}$ a member of itself? That is, is the set of all Russell sets itself a Russell set? There can be only two possible answers to this question: yes or no.

Suppose the answer is yes. Then $\mathbf{R}$ is a member of $\mathbf{R}$. To become a member, $\mathbf{R}$ must have met the membership criterion that is italicized two paragraphs above: $\mathbf{R}$ is not a member of itself. Hence, if $\mathbf{R}$ *is* a member of $\mathbf{R}$, then $\mathbf{R}$ cannot be a member of

R. This clear contradiction rules out the possibility of a "yes" answer to the fatal question.

But what if the answer is no and **R** is *not* a member of **R**? Then **R** is certainly not a member of itself and, like our set of teaspoons, meets the membership criterion for admission to **R**. So, if **R** is not a member of **R,** it must automatically become a member of **R**. Again we are staring a contradiction in the face.

To Russell it all should have been so simple. Yet somehow "each alternative leads to its opposite and there is a contradiction." He stood baffled by the "very peculiar class" he had created with reasoning that "had hitherto seemed adequate."[25] It is what we now call Russell's paradox.

It may be helpful to present a somewhat more concrete illustration of the logical twists of his argument. Suppose a noted art connoisseur decides to classify all the world's paintings into one of two mutually exclusive categories. One category, admittedly quite rare, consists of all paintings which include an image of themselves in the scene depicted upon the canvas. As an example, we might paint a picture, titled *Interior,* of a room and its furnishings—flowing draperies, a statue, a grand piano—which includes, hanging over the piano, a tiny depiction of the painting *Interior.* Thus our canvas would include an image of itself.

The other category, far more common, consists of all paintings that do *not* include an image of themselves. We shall call pictures in this category "Russell paintings." The *Mona Lisa,* for instance, is a Russell painting because it does not show within it a small framed picture of the *Mona Lisa.*

Suppose further that our art connoisseur mounts a huge exhibit that includes all the world's Russell paintings. After much effort, these are collected and hung upon the walls of an enormous hall. Proud of his accomplishment, the connoisseur hires an artist to paint a picture of the hall and its contents.

When the painting is completed, the artist properly titles it *All the World's Russell Paintings* and presents it to the connoisseur. He examines her work carefully and detects a minor flaw: There on the canvas, next to a representation of the *Mona Lisa,* is a painted image of *All the World's Russell Paintings.* This means that *All the World's Russell Paintings* is a picture including an image of itself and thus is *not* a Russell painting. Consequently it does not belong in the exhibit and certainly should not be shown hanging upon the wall. He asks the artist to brush it out.

She does so and again presents the picture to the connoisseur. After some scrutiny, he realizes that there is a new problem: The painting *All the World's Russell Paintings* now does not include an image of itself and so is a Russell painting that belongs in the exhibit. Consequently it should be depicted as hanging somewhere on the wall lest the work not include *all* Russell paintings. The connoisseur therefore recalls the artist and asks her to dab in a little image of *All the World's Russell Paintings.*

But once the image is added, we are back to square one. The image must then be removed, after which it must be restored, then removed, and so on. Sooner or later (it is hoped) the artist and connoisseur will realize that something is amiss: They have stumbled upon Russell's paradox.

All of this might seem utterly irrelevant. But recall that the object of Russell's work was to build all of mathematics upon the unshakable foundation of logic. His paradox jeopardized such a program. Just as the occupant of the penthouse suite should feel uneasy upon learning of a crack in the basement, so should mathematicians feel uneasy knowing that, at the very foundation of their subject, there lies a crack in the logic. It suggests that the whole mathematical enterprise, like the apartment tower, could come tumbling down at any time.

Needless to say, Russell was shaken by the existence of his paradox. "I felt about the contradictions," he wrote, "much as an earnest Catholic must feel about wicked Popes."[26] Others were similarly dismayed, as is evident in the famous exchange between Russell and the logician Gottlob Frege (1848–1925). The latter had published *Grundgesetze der Arithmetik*, a huge work aimed at exploring the foundations of arithmetic. In it Frege had worked with sets in the same naive and cavalier manner that had led Russell to his paradox. Russell communicated his example to Frege, who immediately recognized it as dealing a death blow to his enterprise. In the second volume of his *Grundgesetze*, which was going to press at the time of Russell's letter, Frege had to face every scholar's greatest nightmare: that his or her work, at the last minute, has been rendered incorrect. With a poignancy matched only by his integrity, Frege wrote: "A scientist can hardly meet with anything more undesirable than to have the foundations give way just as the work is finished. I was put in this position by a letter from Mr. Bertrand Russell when the work was nearly through the press."[27]

The statement of the paradox was clear, but its resolution was not. After years of unsuccessful attempts, logicians eventually tried to legislate it away by stipulating that a set that contained itself as a member is really not a set. By means of such logical tactics, and some carefully crafted definitions, such classes were proclaimed to be illegitimate.

The reasonableness of this approach can perhaps be illustrated by our fable about the paintings. Is it even permissible to talk about a painting that contains a representation of itself? If *All the World's Russell Paintings* contained an image of itself, then closer scrutiny of that image—perhaps with the aid of a magnifying glass—would reveal a tiny image of *All the World's Russell Paintings*. Within this would have to be yet a tinier version of *All the World's Russell Paintings*. And so it would go, forever, like the endless reflections in a clothing store mirror. A painting with such an infinite regress could never really be painted.

In a crude sense, this illustrates Russell's attempted resolution of the paradox. He wrote that "whatever involves all members of a collection must not itself be a member of the collection."[28] Consequently, the self-referential nature of membership in the Russell set was illegitimate. The Russell set is not a set at all.

This solution, which required some excruciating convolutions of thought, seemed cumbersome and artificial. Russell spoke of it as "theories which might be true but were not beautiful."[29] If nothing else, it transferred the study of sets from the naive, pre-Russellian domain into a less intuitive realm.

Bertrand Russell
(Courtesy of The Bertrand Russell Archives, McMaster University)

For those mathematicians indifferent to foundational questions, the entire matter seemed to require more thought than it was worth. And Russell came to believe that his ultimate reduction of mathematics to logic was less satisfactory than he had, in his youthful optimism, envisioned.

The intellectual strain and its disappointing conclusion exacted a terrible toll. Russell recalled how he thereafter "turned aside from mathematical logic with a kind of nausea."[30] There were more thoughts of suicide, although he chose to abstain because, as he observed, he would surely live to regret it. Gradually the disappointment passed and, as we have seen, he kept up the good fight for another two-thirds of a century.

It is difficult, in the final analysis, to summarize his long life. He was an irresistible intellectual force and the twentieth century's great curmudgeon. He despaired about the human condition yet fought to improve it. He was labeled a scoundrel as often as he was proclaimed a hero. But even his worst enemies could not deny that the man had the courage of his convictions. As his grandmother had advised, he did not follow a multitude to do evil.

We leave him with the last word. In a 1925 essay, "What I Believe," Bertrand Russell provided a clue to what sustained him through his long and turbulent life. "Happiness," wrote the great skeptic,

is none the less true happiness because it must come to an end, nor do thought and love lose their value because they are not everlasting. . . . Even if the open windows of [reason] at first make us shiver after the cosy indoor warmth of traditional humanizing myths, in the end the fresh air brings vigour, and the great spaces have a splendour of their own.[31]

Spherical Surface

$S = 4\pi r^2$

The sphere is simplicity itself. No three-dimensional body can be more easily defined, and none exhibits more perfect symmetry. It is indisputably *pure*.

Like countless people before and since, the philosopher Plato celebrated its perfection. He asserted that the creator of the universe "turned it into a rounded, spherical shape, with the extremes equidistant in all directions from the centre, a figure that had the greatest degree of completeness and uniformity, as he judged uniformity to be incalculably superior to its opposite."[1] In more recent times Cézanne perceived this superiority and advised artists to "Treat nature in terms of the cylinder, the sphere, the cone, all in perspective."[2] His painterly eye saw spheres everywhere, overhead and underfoot. Indeed, but for a handful of astronauts, the human family has trod upon a very large sphere throughout all of its history.

Even as the sphere is ubiquitous, it also has an undeniable beauty, an elegance arising from its inherent simplicity and setting it apart from all other shapes. No solid body has quite the sphere's claim to our attention.

Such musings aside, the sphere is really a mathematical entity. Strictly speaking, a *sphere* is defined as the set of all points in space a given distance from a fixed point. The given distance, of course, is the *radius* and the fixed point is the *center*. Euclid, however, adopted a more dynamic viewpoint when he defined a sphere in these terms: "When, the diameter of a semicircle remaining fixed, the semicircle is carried round and restored again to the same position from which it began to be moved, the figure so comprehended is a sphere."[3]

The idea of a sphere's being swept out by a revolving semicircle, as illustrated in Figure S.1, brings an exhilarating sense of action. Of course, it also suggests that a tangible semicircle must be moved physically through space. Modern mathemati-

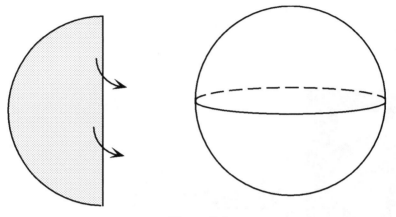

Figure S.1

cians prefer definitions rooted in pure logic to those depending on physical motion. Nonetheless, Euclid's concept of generating a sphere by revolution was central to determining its surface, and the mathematician who put all the pieces together was the incomparable Archimedes of Syracuse.

The objective of the chapter is simply this: to follow Archimedes in finding the surface area of the sphere. It is a question that led him quickly into very deep waters. Before plunging in ourselves, we shall derive the surface area of a less challenging three-dimensional figure, the cylinder.

The standard derivation, illustrated in Figure S.2, cuts the cylinder vertically, unrolls it, and flattens it out (note that we assume the cylinder has neither top nor bottom). The result is a rectangle whose height is the height of the original cylinder and whose width is the circumference of the circle at the cylinder's base. This circumference, as we saw in Chapter C, is $\pi D = 2\pi r$. Therefore

surface area of cylinder = area of rectangle = $b \times h = (2\pi r) \times h = 2\pi rh$

The ease of this argument shows that cylinders, though curvy, are not *impossibly* curvy.

Finding the area of the curved surface of a sphere, however, is far from trivial. For one thing, it is unclear where to begin. Taking a cue from the cylinder, we might try to cut and unroll a sphere, but the outcome is not a simple or recognizable figure. Or we might try to paste a host of little squares upon the sphere's surface to get its area, but they never quite cover it all. Using square measures for the surface of a sphere is like comparing apples and oranges.

These difficulties were not enough to keep Archimedes from probing its deepest secrets. In Chapter C we noted that he achieved a succession of mathematical triumphs the likes of which had not been seen before and rarely since. The greatest of these, in his own opinion and that of succeeding generations, was the determination of the surface area and volume of the sphere, wonderful discoveries from his master-

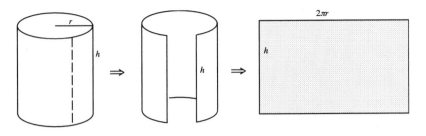

Figure S.2

piece *On the Sphere and the Cylinder*. As we shall see, the geometry gets a bit involved, but the outcome is worth a little intellectual exertion.

It is a truism in mathematics that a difficult problem can often be approached through a series of simpler subproblems. (Actually this is not a bad lesson for dealing with life generally.) This truism was not lost upon Archimedes. Rather than attacking spherical surface directly, he relied upon the properties of two more accessible solids: cones and conical frusta. Following in his footsteps, we derive their surface areas.

Suppose we have a cone as depicted in Figure S.3. The circle at its base has radius r and the length of the straight line along the conical surface from vertex to base—the so-called *slant height*—is s.

To determine the cone's area, we slice it from base to vertex as indicated, then flatten out the resulting surface to get a portion, technically called a *sector*, of a circle. Notice that the slant height s of the original cone has become the radius of this sector.

We now complete the circle of which this sector is a part, as shown in Figure S.4. It is clear that the ratio of the area of the sector to the circle's total area equals the ratio of the arc length along the rim of the sector to the circle's total circumference. In other words,

$$\frac{\text{area of sector}}{\text{area of circle}} = \frac{\text{arc length of sector}}{\text{circumference of circle}}$$

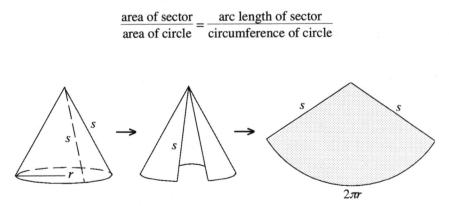

Figure S.3

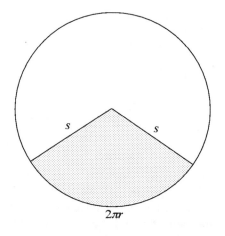

Figure S.4

For instance, if the sector accounted for one-third of the area of the restored circle, then its arc length would likewise be one-third of the circumference of the restored circle.

Of course, the restored circle with radius s has area πs^2 and circumference $2\pi s$. And we see from Figure S.3 that the arc length of the sector is just the circumference of the circle at the base of the original cone: $2\pi r$. Assembling this information, we get

$$\frac{\text{area of sector}}{\pi s^2} = \frac{2\pi r}{2\pi s} = \frac{r}{s}$$

and a cross-multiplication yields

$$\text{area of sector} = \frac{r}{s} \times \pi s^2 = \pi rs$$

Because the area of the flattened sector is exactly the surface area of the original cone, we have proved

FORMULA A: Surface area of cone $= \pi rs$, where r is the radius and s is the slant height.

The second surface Archimedes needed was that of a frustum of a cone. A *frustum* is the lower solid remaining when a cone is sliced by a plane parallel to the base and the top is removed, as shown in Figure S.5. Letting r be the radius of the circle at the top of the frustum, R be the radius of that on the bottom, and s be the slant height of the frustum—that is, the length of the line running perpendicularly down the conical surface from upper to lower circle—we must determine the frustum's surface area (again, not including its top or bottom).

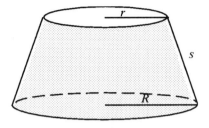

Figure S.5

The most natural way to proceed is to restore the missing top of the cone, and then use formula A to find the surface area of the large, restored cone and that of the smaller one. The difference between these will be the surface area of the frustum.

For notational ease, we call the slant height of the upper part t, as indicated in Figure S.6. Because the upper cone has radius r and slant height t, its surface area is $\pi r t$ by formula A. For the large restored cone, the base radius is R and the slant height is $s + t$, the sum of the slant heights of the upper cone and the frustum. Therefore its surface is $\pi R(s + t)$. It follows that

$$\text{surface area of frustum} = \text{surface of restored cone} - \text{surface of upper cone}$$

$$= \pi R(s + t) - \pi r t$$

$$= \pi R s + \pi R t - \pi r t = \pi[Rs + (Rt - rt)]$$

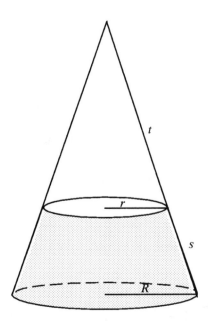

Figure S.6

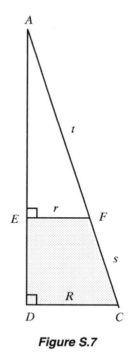

Figure S.7

Unfortunately, this expression leaves something to be desired, for it requires knowledge of the length t. We much prefer a formula involving only R, r, and s—dimensions of the original frustum—and not this ghost quantity t that measures a part of the cone that long since has been discarded. Our expression, though correct, remains the source of some "frustation."

The situation can be remedied by introducing similar triangles. Suppose we slice a vertical plane through the restored cone, thereby generating Figure S.7. Clearly the upper right triangle AEF is similar to the large right triangle ADC because both contain a right angle and $\angle DAC$. By similarity, the corresponding sides are proportional and, in particular, the ratio of hypotenuse to horizontal side is the same in both triangles. Thus $t/r = (s + t)/R$. Cross-multiplying and simplifying algebraically yields

$$Rt = r(s + t) = rs + rt \quad \text{or} \quad Rt - rt = rs$$

We then substitute this into the bracketed expression from our earlier formula for the frustum's area to get

$$\text{surface area of frustum} = \pi[Rs + (Rt - rt)] = \pi[Rs + rs] = \pi s(R + r)$$

To summarize, we have proved

FORMULA B: Surface area of frustum $= \pi s(R + r)$, where the frustum has upper radius r, lower radius R, and slant height s.

In words, this says the surface area of the frustum of a cone is the product of π, the slant height, and the sum of radii of the circles forming its bases.

The preliminaries are now complete, but spherical surface is still nowhere in sight. In fact, at this point Archimedes unexpectedly turned his attention not to the three-dimensional sphere but to the two-dimensional circle. Hold onto your hats.

Within a circle of radius r and diameter AA', he inscribed a regular polygon with an even number of sides, each of length x. For purposes of illustration in Figure S.8, we have used regular octagon $ABCDA'D'C'B'$, but the reasoning will carry over to any even-sided polygon. Archimedes drew the vertical lines BB', CC', and DD' intersecting diameter AA' at F, G, and H; the dotted lines $B'C$ and $C'D$ intersecting the diameter at K and L; and the seemingly unimportant line $A'B$, whose length we shall denote by y. With this, his diagram was cut into a forbidding array of large and small triangles.

There are two immediate consequences of these constructions. One is that segments BF and $B'F$ are the same length—call it b; that CG and $C'G$ are the same length—call it c; and that DH and $D'H$ are the same length—call it d.

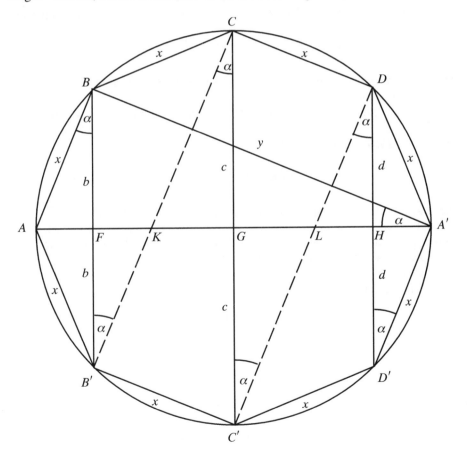

Figure S.8

The second requires us to invoke a result from Book III of Euclid's *Elements* to the effect that angles in circles that intersect equal arcs are themselves equal. Because we began with a *regular* polygon, the small arcs into which the circle is divided by the polygonal sides are equal, so all angles inscribed in the circle and intersecting these arcs have the same measure. Thus $\angle BA'A = \angle ABB'$ because they intercept equal arcs AB and AB', respectively; for the same reason, $\angle ABB' = \angle BB'C = \angle B'CC'$, and so on. In Figure S.8 each of these equal angles is denoted by α.

We now follow Archimedes in building a string of proportions. Note that $\triangle ABA'$ and $\triangle AFB$ are similar because both share $\angle BAA'$ and both have an angle of size α. By proportionality of corresponding sides we conclude

$$\overline{AF}/\overline{BF} = \overline{AB}/\overline{A'B} \text{ or simply } \frac{\overline{AF}}{b} = \frac{x}{y}$$

Cross-multiplication yields $xb = (\overline{AF})y$, a result we shall store for later use.

Next observe that $\triangle AFB$ is similar to $\triangle KFB'$ because both contain an angle of size α and because $\angle AFB$ and $\angle KFB'$ are vertical angles. This yields the proportion

$$\overline{FK}/\overline{B'F} = \overline{AF}/\overline{BF} \quad \text{or} \quad \frac{\overline{FK}}{b} = \frac{\overline{AF}}{b} = \frac{x}{y}$$

where the last equality simply repeats the formula from the previous paragraph. Cross-multiplication gives $xb = (\overline{FK})y$.

We continue marching along the circle in pursuit of similar triangles. Next in line are $\triangle KFB'$ and $\triangle KGC$, which share an angle of size α and which contain the vertical angles $\angle FKB'$ and $\angle GKC$. Thus,

$$\overline{KG}/\overline{CG} = \overline{FK}/\overline{B'F} \quad \text{or} \quad \frac{\overline{KG}}{c} = \frac{\overline{FK}}{b} = \frac{x}{y}$$

where again the last equality comes from the previous paragraph. Hence $xc = (\overline{KG})y$.

And on we go, concluding that $\triangle KGC$ is similar to $\triangle LGC'$, from which it follows, exactly as above, that $xc = (\overline{GL})y$. Likewise, the similarity of $\triangle LGC'$ and $\triangle LHD$ implies that $xd = (\overline{LH})y$, and the similarity of $\triangle LHD$ and $\triangle A'HD'$ gives $xd = (\overline{HA'})y$.

So what is to be done with this batch of equations? Archimedes added them:

$$xb = (\overline{AF})y$$
$$xb = (\overline{FK})y$$
$$xc = (\overline{KG})y$$
$$xc = (\overline{GL})y$$
$$xd = (\overline{LH})y$$
$$+ \, xd = (\overline{HA'})y$$

$$\rule{11cm}{0.4pt}$$

$$xb + xb + xc + xc + xd + xd = (\overline{AF} + \overline{FK} + \overline{KG} + \overline{GL} + \overline{LH} + \overline{HA'})y$$

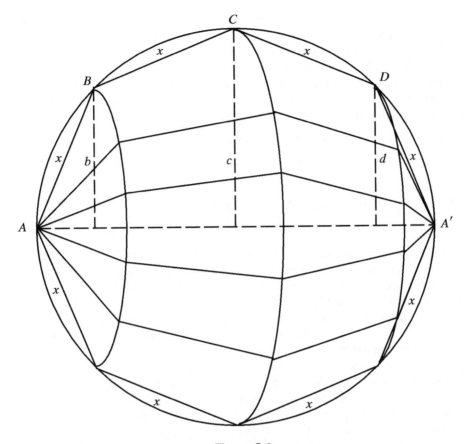

Figure S.9

This reduces to the much simpler

$$x[2b + 2c + 2d] = (\overline{AA'})y$$

because the segments on the right-hand side collectively form the circle's diameter. Because the circle's radius was given to be r, we know $\overline{AA'} = 2r$. Thus we have proved:

$$x[2b + 2c + 2d] = 2ry \qquad (*)$$

Although it is yet unclear how Archimedes will use this, the relationship labeled (*) plays a pivotal role in what follows.

In the next step we at last encounter a sphere. Archimedes revolved the entire configuration of Figure S.8 about the horizontal axis AA'. As Euclid's definition promised, the revolving circle sweeps out a sphere; simultaneously the revolving polygon generates a solid consisting of a series of conical frusta with cones attached on either end, as shown in Figure S.9.

It is important to note that the slant height of each cone and frustum is x, the length of each side of the original inscribed regular polygon.

We now determine the surface area of this solid. For the cone on the left, with slant height x and base radius b, the surface area is πxb by formula A. The left-hand frustum has slant height x, upper radius b, and lower radius c, so formula B tells us its surface is $\pi x(b + c)$. Likewise the right-hand frustum has surface $\pi x(c + d)$, and the cone on the right has surface πxd. Combining results, we have

$$\text{surface of inscribed solid} = \pi xb + \pi x(b + c) + \pi x(c + d) + \pi xd$$
$$= \pi x[b + (b + c) + (c + d) + d] = \pi x[2b + 2c + 2d]$$

Here, as if by magic, we have reached an expression containing (*) above. Substitution brings us to the critical point in this long argument:

$$\text{surface area of inscribed solid} = \pi x[2b + 2c + 2d] = \pi(2ry)$$

It now becomes evident why Archimedes introduced the mystery line $A'B$: Its length y figures in the surface area of the solid within the sphere. It is likewise apparent why he needed a regular polygon with an *even* number of sides. In this way each internal radius (b, c, and d in our figure) is shared by two solids. Had Archimedes begun with an odd number of sides, he would not have gotten cones on either end; consequently, one radius would fail to be part of both a frustum and adjoining cone, and this would rule out use of equation (*).

In any case, we have determined the surface area of the inscribed solid, not that of the sphere itself. But the former serves as an approximation of the latter, an approximation that gets better as the number of sides of the regular polygon grows. Rather than inscribing an octagon, we could use a decagon, or a 20-gon, or a 20,000,000-gon. Regardless of the number of sides, the surface area of the inscribed solid, by the argument above, will be $\pi(2ry)$. At the same time, the solid's surface will approach that of the sphere. We thus apply the notion of "limit" from Chapter D to see that:

$$\text{surface area of sphere} = \lim(\text{surface area of inscribed solids}) = \lim \pi(2ry)$$

As the number of sides of the polygon increases indefinitely, there is no change in the value of r, the sphere's radius. But y was the length of segment $A'B$, and this does change. Clearly, as the number of sides of the polygon grows, the point labeled B in Figure S.8 descends along the arc of the circle toward A, so segment $A'B$ approaches diameter AA'. In other words,

$$\lim y = \lim \overline{A'B} = \overline{A'A} = 2r$$

With this we finally reach the desired end:

$$\text{surface area of sphere} = \lim [\pi(2ry)] = 2\pi r[\lim y] = 2\pi r(2r) = 4\pi r^2$$

This conclusion is as simple to state as it was complicated to prove. ■

In his treatise *On the Sphere and the Cylinder*, Archimedes stated the theorem differently. Because he was working almost 2,000 years before the coming of algebraic notation, a formula like $4\pi r^2$ would have held no meaning. Instead he presented the theorem in words with a hint of poetry: "The surface of any sphere is four times the greatest circle in it."[4] This, of course, agrees with our version above because the "greatest circle" in a sphere is that formed by slicing through the sphere along a diameter. This circular cross-section has radius r and area πr^2, so when Archimedes said the area was four times as great, he was saying that the spherical surface is $4\pi r^2$. Whether expressed as a formula or a sentence, this is truly a wonderful piece of reasoning.

For the sake of historical accuracy, we must add a few disclaimers. Our argument followed the route taken by Archimedes but with some significant variations. First, as noted, he proceeded in a purely geometric, as opposed to algebraic, fashion. Second, he did not use limits. After the argument's critical point—the determination of the surface area of the inscribed, approximating solid—we simply let the number of sides of the regular polygon tend to infinity, took a limit, and were done.

But Archimedes, without the limit concept or the algebraic notation underlying it, employed the proof technique called double *reductio ad absurdum* that we saw in our discussion of Euclid's work in Chapter G. That is, he first proved that it was impossible for the surface area of the sphere to be *more* than four times its greatest circle. He then turned around and showed that it was impossible for the spherical surface to be *less* than four times its greatest circle. Only after having dispensed with these two alternatives could he conclude that the surface of the sphere was exactly four times its greatest circle, not a shade more, not a shade less.

We should in no way condemn Archimedes for the indirectness of his reasoning. In his skilled hands double *reductio ad absurdum* was sufficient to establish this and many other significant results of geometry, and mathematicians more than 1,500 years later were still using the technique. He did a brilliant job with the tools at his disposal. Only with the introduction of algebraic notation and limits could mathematicians take the radical shortcut used above.

So, this is the great theorem from *On the Sphere and the Cylinder*. Elsewhere in that work Archimedes gave a different version of the result, one that explains the significance of the title. He wrote, "Any cylinder having its base equal to the greatest circle of those in the sphere, and height equal to the diameter of the sphere is . . . half as large again as the surface of the sphere."[5] By "half as large again" he meant that

$$\text{cylindrical surface} = \text{spherical surface} + \frac{1}{2}(\text{spherical surface})$$

Here Archimedes had related a cylinder to the sphere fitting snugly within it (see Figure S.10). But is this statement equivalent to the previous one? The answer, of course, is "yes," and we shall quickly see why.

Earlier in this chapter we proved that the surface area of a cylinder is $2\pi rh$, and because the sphere fits into the cylinder, the height of the latter is just the diameter of

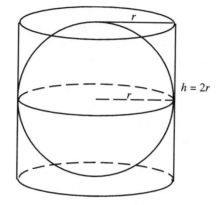

Figure S.10

the sphere; that is, $h = 2r$. Hence the lateral surface of this cylinder is $2\pi r(2r) = 4\pi r^2$.

But this time when Archimedes spoke of the cylinder's area, he was including the top and bottom as well. The area of the cylinder's circular top is πr^2 and the bottom is the same. Hence the overall surface of the cylinder is

$$\text{lateral surface} + \text{top surface} + \text{bottom surface} = 4\pi r^2 + \pi r^2 + \pi r^2 = 6\pi r^2$$

Archimedes stated that this cylindrical surface is "half as large again as the surface of the sphere." Letting S stand for the sphere's surface area, we thus have

$$6\pi r^2 = \text{cylindrical surface} = S + \frac{1}{2}S$$

We double both sides to get $12\pi r^2 = 2S + S = 3S$, and so $S = (1/3) \times 12\pi r^2 = 4\pi r^2$, exactly as before.

Archimedes was quite intrigued by this cylinder/sphere relationship and was justifiably proud to have discovered it. According to legend, he asked that a figure of a sphere within a cylinder be carved upon his tombstone to serve as a reminder of this great geometrical truth. It was to be his monument.

We conclude with an observation about the past. It is easy, amid the scientific and technological advances of the modern age, to feel a certain intellectual superiority over those generations who came before. Aristotle, after all, never earned a Ph.D. and Euclid didn't win the Nobel Prize. We sit back, click on the television, and pity our intellectually limited predecessors.

This chapter should stifle any such feelings. Surely the mathematics we have just seen dispels the notion that all the world's smart people are alive today. Under the penetrating gaze of Archimedes more than 20 centuries ago, the mystery of spherical surface was resolved forever.

Trisection

$$\angle ABD = \frac{1}{3}\angle ABC$$

For longer than anyone can remember there has been a fascination with brave heroes undertaking impossible quests. From the Holy Grail to the buried treasure of Captain Kidd, from the Northwest Passage to the fountain of youth, adventurers have marched away with hopes high. Many returned broken and disappointed. Some did not return at all. A few, against all odds, succeeded: Jason found the Golden Fleece, Curie isolated radium, Hillary and Tenzing reached the top of Everest. This is the stuff of legend, for tales of such perseverance and courage have a powerful grip on us all.

Mathematics certainly has had its share of quests, both successful and unsuccessful, although these are undertaken in the rarefied atmosphere of pure reason and not in the rarefied atmosphere of the Himalayas. Among them, none is more famous than the millennia-old search for angle trisection.

The origins of this story, like so many in mathematics, go back to the Greek geometers. The challenge was simply put: Start with an arbitrary angle and divide it exactly into thirds. This task seems fairly straightforward, but at the outset we should clarify the rules.

First, we are limited to using the tools of geometry—the compass and unmarked straightedge discussed in Chapter G. Trisections that require other tools, even if ingenious, do not solve the problem. Indeed, Greek geometers performed trisections by introducing auxiliary curves like the quadratrix of Hippias or the spiral of Archimedes, but these curves were not themselves constructible with compass and straightedge and thus violated the rules of the game. It is rather like reaching the top of Everest by helicopter: It achieves the end by an unacceptable means. For a legitimate trisection, only compass and straightedge need apply.

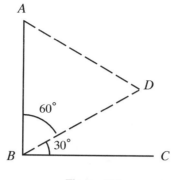

Figure T.1

The second rule is that the construction must require only a finite number of steps. There must be an end to it. An "infinite construction," even if it has trisection as a limiting outcome, is no good. Construction that goes on forever may be the norm for interstate highways, but it is impermissible in geometry.

Finally we must devise a procedure to trisect *any* angle. Trisecting a particular angle, or even a thousand particular angles, is insufficient. If our solution is not general, it is not a solution.

This last point is illustrated in Figure T.1. Suppose with compass and straightedge we construct AB perpendicular to BC (a simple procedure). With AB as base, next construct equilateral triangle ABD. As we discussed in Chapter G, this was the very first proposition in the *Elements* and so is perfectly legitimate. Now $\angle ABD$ contains 60° and $\angle ABC$ contains 90°, so $\angle DBC$ has measure 30° = (1/3) ($\angle ABC$). With compass and straightedge we have therefore trisected a right angle.

Is this cause for celebration? Not especially, because the trisection of a right angle was not the objective. It is the *general* angle that must be our target, and the procedure above is certainly not general.

A phenomenon that may have encouraged angle trisectors in their quest is that two apparently related constructions *can* be performed with compass and straightedge. One is the bisection of any angle and the other is the trisection of any line segment. We digress a moment to see how these are done.

First, suppose we are given an arbitrary angle $\angle ABC$ as shown in Figure T.2 and wish to bisect it with compass and straightedge. The procedure we use appears as proposition 9 of Book I of the *Elements*. Begin by choosing any point D on segment AB. With compass centered at B and having radius BD, draw an arc cutting BC at E so that $\overline{BD} = \overline{BE}$. Use the straightedge to draw DE, and upon DE construct equilateral triangle DEF. Finally, draw segment BF.

The theory of triangle congruence proves that BF bisects $\angle ABC$, for $\overline{BD} = \overline{BE}$ by construction; $\overline{DF} = \overline{EF}$ because $\triangle DEF$ is equilateral; and $\overline{BF} = \overline{BF}$. We conclude that $\triangle BDF$ is congruent to $\triangle BEF$ by *SSS,* and so $\angle ABF$ has the same measure as $\angle CBF$. In other words, $\angle ABC$ has been split in half.

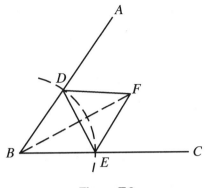

Figure T.2

We note that we have bisected a *general angle* using *compass and straightedge* in a *finite number of steps*, thereby satisfying all of our rules. Angle bisection is obviously quite elementary.

It is also easy to quadrisect an angle, that is, divide it into four equal parts. We need only bisect ∠*CBF* and ∠*ABF* by repeating the preceding construction in order to get perfect fourths. Bisecting each of these would produce perfect eighths, and so on. Clearly the division of a general angle into 2^n equal parts presents no difficulty. Of course, this is of no help whatever in dividing the angle into thirds.

The other related construction is the compass-and-straightedge trisection of a general line segment. Again we look to Euclid, who described the following procedure as proposition 9 of Book VI of the *Elements*.

Start with arbitrary line segment *AB*, which we wish to trisect (see Figure T.3). Draw any other line *AC* emanating from *A*, and choose *D* to be any point on *AC*. With the compass, construct along *AC* the segments *DE* and *EF* both equal in length to *AD*, thereby making *AD* one-third as long as *AF*.

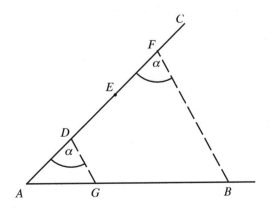

Figure T.3

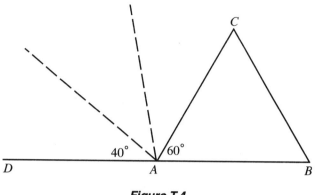

Figure T.4

Next draw BF, forming $\angle AFB$ whose measure we designate by α. With the compass and straightedge, construct $\angle ADG$ also of measure α (a construction Euclid explained as proposition 23 of Book I). This means that $\triangle ADG$ and $\triangle AFB$ are similar because both contain an angle of size α and share the common vertex angle at A. From similarity follows proportionality of corresponding sides. In particular, we have

$$\overline{AG}/\overline{AB} = \overline{AD}/\overline{AF} = \frac{1}{3}$$

by our observation above. So, segment AG is a third as long as AB. We have thereby trisected a *general segment* using *compass and straightedge* in a *finite number of steps*.

If we can bisect angles and trisect lines, it seems reasonable to expect that we can trisect angles. So it must have appeared to the ancient Greeks, and so it appeared to countless mathematicians over the centuries.

There may have been one other fact giving hope to trisectors: The compass and straightedge are capable of some surprisingly sophisticated constructions. No one would be shocked to learn that an equilateral triangle or a square is constructible, but it is far less obvious that compass and straightedge are adequate to construct a regular pentagon, a procedure Euclid described in the fourth book of the *Elements*. Moreover, we can construct a regular hexagon, octagon, decagon, dodecagaon (12 sides), and even pentadecagon (15 sides), the last of which appears as the final proposition of Book IV.

If compass and straightedge have this much power, we might optimistically undertake the construction of, for instance, a regular nonagon (9 sides). A natural place to begin would be to construct equilateral triangle ABC and extend one of the sides to D, as shown in Figure T.4. Then the measure of $\angle DAC$ is $180° − 60° = 120°$. Now, *if* we could trisect $\angle DAC$, we would have constructed an angle of $(1/3)(120°) = 40°$, which is exactly one-ninth of the way around the $360°$ circle. Transferring this $40°$ to the center of a circle and replicating it nine times would produce a regular nonagon, as shown in Figure T.5.

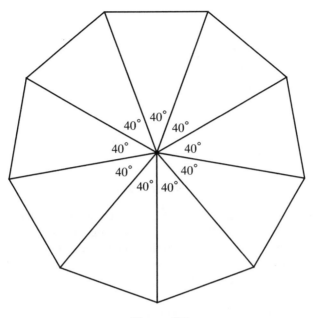

Figure T.5

This construction, of course, hinges on the *if* clause. Undoubtedly the desire to construct a regular nonagon was another impetus in the quest for compass-and-straightedge trisections.

At this point, we should take a look at a pair of "near misses"; that is, procedures that split a general angle into three equal parts but only by violating one or another of the stated rules.

The first is a clever argument attributed to Archimedes. The procedure uses a well-known result, namely that the exterior angle of a triangle equals the sum of the two opposite and interior angles. To prove it, simply extend side AB to D, generating exterior angle DBC in Figure T.6. We know that $\alpha + \beta + \gamma = 180°$ because they are the angles of a triangle; and we know that $\angle DBC + \beta = 180°$ because AD is a straight line. Hence $\alpha + \beta + \gamma = \angle DBC + \beta$ and subtracting β from each side yields $\alpha + \gamma = \angle DBC$, as required.

We now move to Archimedes' trisection of a general angle AOC, as shown in Figure T.7. With center O and any radius r, construct a semicircle and continue segment CO to the semicircle at B and beyond.

It is critical to extend this line just the right distance in a leftward direction. We do the following: Take the straightedge and place one end at A and the other at a point D on the extended line, so that the distance along the straightedge from D to E (where E is the intersection of the semicircle and line AD) is the semicircle's radius. In other words, construct AD so that $\overline{ED} = r$. We claim that $\angle ADC$ thus formed is exactly one-third as large as the original $\angle AOC$.

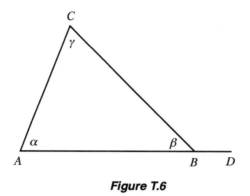

Figure T.6

To prove this, let the measure of ∠*ADC* be α. Draw *EO*, a radius of the semicircle, to create Δ *DEO* having $\overline{ED} = \overline{EO} = r$. This triangle is isosceles, so ∠*EOD* = α as well. Next observe that ∠*AEO*, an exterior angle of Δ *DEO*, has measure equal to the sum of the two opposite interior angles. That is, ∠*AEO* = α + α = 2α. But Δ *EOA* is also isosceles because two of its sides are radii, and so ∠*EAO* =∠*AEO* = 2α.

We now draw the critical conclusion: ∠*AOC*, as an exterior angle of Δ *AOD*, equals the sum of the two opposite interior angles. Hence

$$\angle AOC = \angle ODA + \angle DAO = \alpha + 2\alpha = 3\alpha$$

This means the original ∠*AOC* is precisely three times as great as ∠*ADC;* equivalently, we have constructed ∠*ADC* to be one-third of the given ∠*AOC.* We then copy an angle of the same size within ∠*AOC,* and trisection would be accomplished using only compass and straightedge.

Or is it? Unfortunately an illegal procedure was employed in the midst of this argument. It occurred when the point *D* was being sought. How, in fact, would we really go about using an unmarked straightedge to locate *D* (and thus *E*)? How do we align a straightedge from *A* to guarantee that segment *ED* has length *r*? One can imagine putting marks on the straightedge and then wiggling it about to get the desired fit, but these are not allowable operations. The straightedge must not be marked; use of it must not require a skilled eye to wiggle and jiggle and thereby esti-

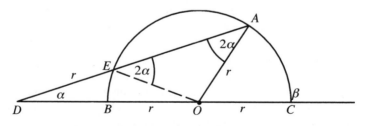

Figure T.7

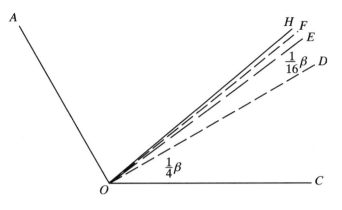

Figure T.8

mate a certain length. This construction, though it certainly trisects the angle, violates the rules of the game.

In all fairness to Archimedes, he recognized the violation. The Greeks even had a term, *verging,* for such wiggling and sliding of the straightedge. So, rather than fault Archimedes for a blunder, we should commend him for a very clever line of reasoning.

Our second near miss also trisects in an illegitimate fashion. Again begin with $\angle AOC$ having measure β, as shown in Figure T.8. By twice bisecting, we get $\angle DOC$ of measure $(1/4)\beta$. If we twice bisect $\angle DOC$, we get an angle of measure

$$\frac{1}{4}\left(\frac{1}{4}\beta\right)=\frac{1}{16}\beta$$

and we copy it as $\angle EOD$. A double bisection of this yields an angle of measure

$$\frac{1}{4}\left(\frac{1}{16}\beta\right)=\frac{1}{64}\beta$$

which is placed as $\angle FOE$.

Continuing indefinitely in this fashion, we build $\angle COH$ of measure

$$\frac{1}{4}\beta+\frac{1}{16}\beta+\frac{1}{64}\beta+\frac{1}{256}\beta+\ldots$$

This is an example of an ***infinite geometric series,*** and we give a quick (although naive) way of evaluating it.

Let S be the sum in question. That is,

$$S=\frac{1}{4}\beta+\frac{1}{16}\beta+\frac{1}{64}\beta+\frac{1}{256}\beta+\ldots$$

We subtract a quarter of S from S to get:

$$S - \frac{1}{4}S = \left(\frac{1}{4}\beta + \frac{1}{16}\beta + \frac{1}{64}\beta + \frac{1}{256}\beta + \ldots\right) - \frac{1}{4}\left(\frac{1}{4}\beta + \frac{1}{16}\beta + \frac{1}{64}\beta + \frac{1}{256}\beta + \ldots\right)$$

$$= \left(\frac{1}{4}\beta + \frac{1}{16}\beta + \frac{1}{64}\beta + \frac{1}{256}\beta + \ldots\right) - \left(\frac{1}{16}\beta + \frac{1}{64}\beta + \frac{1}{256}\beta + \frac{1}{1024}\beta \ldots\right)$$

$$= \frac{1}{4}\beta + \frac{1}{16}\beta + \frac{1}{64}\beta + \frac{1}{256}\beta + \ldots - \frac{1}{16}\beta - \frac{1}{64}\beta - \frac{1}{256}\beta - \frac{1}{1024}\beta \ldots$$

$$= \frac{1}{4}\beta$$

because all but one of the terms on the right cancels. Thus we have

$$S - \frac{1}{4}S = \frac{1}{4}\beta$$

and so

$$\frac{3}{4}S = \frac{1}{4}\beta \rightarrow S = \frac{4}{3} \times \frac{1}{4}\beta = \frac{1}{3}\beta$$

In words, the measure of $\angle COH$ (that is, S) is one third of the measure of the original angle AOC (that is, β). The trisection is complete.

Here the flaw is self-evident: We needed an infinitude of constructions. It is true that the more double bisections we do, the closer we get to a perfect trisection. We can carry the procedure far enough to make it accurate to within a fraction of a degree. But the trisection challenge requires *exact*, not approximate, division into thirds. For this procedure to yield an exact trisection would require infinitely many constructions, and this not only exceeds the stipulated rules but exceeds our finite life spans. It is a process we could never finish.

Such promising attempts notwithstanding, the trisection problem remained unsolved during classical times. By the fourth century A.D., Pappus (whom we met in Chapter I praising the intellect of bees) reported that, "When ancient geometers desired to divide a given rectilinear angle into three equal parts, they were baffled."[1]

The befuddlement persisted through the Renaissance and into the modern era. With each passing century, with each failed attempt, the trisection problem grew in stature. Like an outlaw with a large bounty on his head, trisection was hotly pursued by a posse of mathematicians. Scholars and pseudo-scholars devised trisection procedures and announced them to the world amid great fanfare. Then, without exception, these unfortunate scholars watched as others found flaws in their reasoning. The flood of incorrect proofs got so bad that the Paris Academy declared in 1775 that it would no longer accept trisection arguments.[2] Anyone bringing a trisection proof, like someone bringing the plague, would be turned away at the door.

This policy reflected what some in the mathematical world had come to believe: that the trisection of the general angle was beyond the capacity of compass and straightedge. No less an authority than René Descartes had implied as much more than a century before, and there was growing suspicion that the lack of a proof indicated not that mathematicians were insufficiently clever but that the solution was simply impossible.[3] However, in 1775 this remained only a suspicion; no one had proved impossibility any more than they had divided an angle into thirds.

The danger of drawing a premature conclusion of impossibility was brought home just two decades after the Paris Academy issued its ban. In 1796 the 18-year-old Gauss proved that a regular 17-sided polygon *could* be constructed with compass and straightedge. This was a bombshell. No one before Gauss had the least idea that such a construction was possible, and if the 17-gon had over the centuries generated less interest than trisection, it was only because it seemed even less likely. Gauss's astonishing discovery showed that the compass and straightedge had hidden powers indeed. If the 17-gon could be constructed, maybe someone with a Gaussian intellect could crack the trisection mystery as well.

The question remained open for another few decades until its definitive resolution by Pierre Laurent Wantzel (1814–1848). Wantzel, a mathematician, engineer, and linguist, attended the École Polytechnique in Paris, one of the great scientific training grounds of the day. As sometimes happens with people of such diverse interests, his attention wandered from subject to subject so that he left no voluminous body of work nor any lasting fame. Even among mathematicians, most could not identify the name of Pierre Wantzel.

His obscurity may also be attributed to his short life, and this in turn may be due to intemperate habits. A colleague remembered Wantzel with these words:

> Ordinarily he worked evenings, not lying down until late; then he read, and took only a few hours of troubled sleep, making alternately wrong use of coffee and opium, and taking his meals at irregular hours until he was married. He put unlimited trust in his constitution, very strong by nature, which he taunted at pleasure by all sorts of abuse. He brought sadness to those who mourn his premature death.[4]

Wantzel's 1837 paper on angle trisection was titled "Research on the Means of Knowing If a Problem of Geometry Can Be Solved with Compass and Straightedge."[5] For so significant and longstanding a problem, the work numbered only seven pages, but these were seven remarkable pages indeed. The details of his reasoning carry us well beyond the scope of this book, but we shall at least provide an outline.

Crucial to Wantzel's proof was the transfer of the problem from the realm of pure geometry into that of algebra and arithmetic. He wished to determine which magnitudes could and could not be constructed with compass and straightedge, and to do this he considered the magnitudes not as geometric segments but as numerical lengths.

Wantzel reasoned that if we can trisect the general angle, we can certainly trisect an angle of 60°. Then, adopting an algebraic viewpoint and invoking a bit of trigonometry, he showed that if a 60° angle could be trisected, the cubic equation $x^3 - 3x - 1 = 0$ must have a constructible solution—that is, a solution whose length can be constructed with compass and straightedge. (Actually, Wantzel used a slightly different yet entirely equivalent form of this equation, but that need not concern us here.)

Wantzel's ingenuity became apparent when he proved that if the cubic above has a constructible solution, then it must also have a *rational* solution. That is, there must be a rational number (as defined in Chapter Q) that satisfies this cubic. The issue was thus transferred to investigating whether there exists a rational solution for $x^3 - 3x - 1 = 0$.

Suppose, for the sake of argument, that there is a fraction c/d solving the cubic. We insist that the fraction be in lowest terms; in other words, the numerator c and the denominator d have no common factors, other than the obvious 1 or −1. Because $x = c/d$ is assumed to solve the cubic, we know that

$$(c/d)^3 - 3(c/d) - 1 = 0$$

Multiplying through by d^3 converts this to $c^3 - 3cd^2 - d^3 = 0$.

We now rewrite this equation in two ways. First, observe that $c^3 - 3cd^2 = d^3$, which is equivalent to $c(c^2 - 3d^2) = d^3$. Obviously, the whole number c is a factor of $c(c^2 - 3d^2)$ on the left and so c is also a factor of the equivalent d^3 on the right. But we insisted that c and d have no common factors. From this it follows that the only way c could divide evenly into d^3 is if $c = 1$ or −1.

Returning to the equation $c^3 - 3cd^2 - d^3 = 0$ and arranging it somewhat differently, we see that $3cd^2 + d^3 = c^3$, or equivalently $d(3cd + d^2) = c^3$. Again, it is clear that d is a factor of the left-hand side, so d must be a factor of c^3 as well. Because d and c are without common factors, this implies that d is either 1 or −1.

To summarize: If c/d is a rational solution in lowest terms to the cubic $x^3 - 3x - 1 = 0$, then $c = \pm 1$ and/or $d = \pm 1$. But then the fraction c/d can only be 1 or −1.

Thus we have dramatically narrowed the search to these two rational options, which can be checked one at a time. If $x = c/d = 1$, we get $x^3 - 3x - 1 = 1 - 3 - 1 = -3 \neq 0$, and so $c/d = 1$ is not a solution to the cubic. Similarly, if $x = c/d = -1$, we substitute to find that $x^3 - 3x - 1 = (-1)^3 - 3(-1) - 1 = -1 + 3 - 1 = 1 \neq 0$, so $c/d = -1$ is not a solution either. As we know, these were the *only* possible rational solutions, and because neither works, we conclude that this cubic is solved by no rational number.

So where are we? We need only assemble this string of implications to establish the impossibility of trisection. The chain of reasoning proceeds as follows:

1. *If* we can trisect the general angle with compass and straightedge,

2. Then we can surely trisect a 60° angle,

3. So we can find a constructible solution for $x^3 - 3x - 1 = 0$,

4. So we can find a rational solution for $x^3 - 3x - 1 = 0$,

5. And this rational solution must be either $c/d = 1$ or $c/d = -1$.

But, as we checked, statement (5) is false, for neither 1 nor −1 solves the cubic $x^3 - 3x - 1 = 0$. We have reached a contradiction. Because statement (1) led inexorably to statement (5), we conclude that statement (1) is equally invalid. In short, invoking our old friend "proof by contradiction," we have resolved the issue that had so troubled generations of mathematicians: Trisection of the general angle with compass and straightedge is impossible.

Wantzel's was certainly not a simple argument; one does not expect simplicity from a problem that had been kicking around for more than 20 centuries. But it was final. He proudly staked his claim for priority by observing, "It seems that it has not yet been rigorously demonstrated that these problems [trisections], so celebrated among the ancients, are not susceptible to a solution by geometric constructions."[6] So in 1837 the matter should have been settled.

It was—for serious mathematicians. But, strangely, a collection of quasi-serious, misguided, or just plain kooky individuals have persisted in looking for trisections anyway. Even today angle trisectors are busy. Each claims to have discovered the magic process by which trisections are possible and thereby to have earned an honored place in the mathematics history books.

All of them are wrong. Wantzel's proof was conclusive. Trisection cannot be achieved. In the words of Underwood Dudley, one might as well try "to find two even integers whose sum is odd."[7] Yet committed trisectors are not easily discouraged. As Robert Yates observed, "Once the virus of this fantastic disease gets into the brain, if proper antiseptics are not immediately applied, the victim begins a vicious circle that leads . . . from one outrage of logic to another."[8]

One explanation for such behavior is a misunderstanding of the word *impossible*. To some people, *impossible* sounds more like a challenge than a conclusion. After all, it was once deemed impossible for humans to fly, or to bridge the Golden Gate, or to reach the moon. Yet each of these impossible challenges was met. And who among us has not heard the ringing declamation that "In America, nothing is impossible!" Never mind that the speakers in such cases are usually politicians or authors of self-help books.

Mathematicians know better. As shown in Chapter J, mathematicians can prove negatives in a final, decisive sense. In this case, impossible means, literally, *impossible*.

So those souls who continue to seek the Holy Grail of trisection should be advised that in 1837 P. L. Wantzel proved that if it is possible to trisect, then it is possible to find a rational solution to a certain equation without rational solutions. The logical impossibility of the latter implies the logical impossibility of the former.

Compass and straightedge trisections do not—*cannot*—exist. Case closed.

Utility

tan $\alpha = a/b$

Mathematics is useful.

It is hard to make a more banal statement, for everyone from seasoned scholar to uneasy mathophobe is aware of the wide-ranging applicability of mathematics to problems of the real world. Year in and year out, this awareness fills thousands of math courses and sells hundreds of thousands of textbooks to those for whom mathematics is an indispensable means to an end. Students bound for engineering, architecture, physics, economics, astronomy, or countless other professions are told, quite rightly, that they must acquire a knowledge of mathematics to succeed in their anticipated careers. When it comes to utility, few human undertakings can compete.

The banality of this observation masks a subtle philosophical question: *Why* does mathematics work so well in its utilitarian role? Pure mathematics, after all, is a web of abstractions, a system of ideas internally consistent and logically beautiful, but ideas nonetheless. Logical consistency does not, in and of itself, guarantee usefulness. The rules of cribbage, for instance, are logically consistent but provide no insight about the orbit of the moon.

Or consider the geometry of Euclid described in Chapter G. Without question it is a stunning example of careful deduction from a set of postulates, but that does not *necessarily* mean that Euclid's propositions describe the geometry of the vacant lot across the street. Nonetheless, with a piece of paper and a bit of Euclid we can sit indoors and calculate lengths and areas of the lot that outdoor measurements will subsequently confirm. There really is no need to go outside; the abstractions of mathematics produce results so accurate that the lot itself is hardly necessary.

Yet Euclidean geometry is not about vacant lots. It is not about physical objects. It is about ideas. What is going on here? Why is it that mathematics so often justifies Lord Kelvin's description of it as "the etherealization of common sense?"[1]

Does nature, as is often said, obey mathematical rules? Such an obedience suggests that the outside world is somehow constrained by mathematical principles. Or do nature and mathematics exhibit parallel but essentially unrelated behavior? Is it just serendipitous that mathematics, with its orderly character, is the perfect language to describe the intrinsic order of the world? Perhaps the rhythms and structures of intangible mathematics simply mimic the rhythms and structures of tangible reality with neither obeying the other.

Beyond these philosophical issues, a more mundane fact should be observed: Many natural phenomena have been successful in resisting mathematical solution. Sometimes mathematicians are just not up to the task. This was the opinion of a skeptical Frederick the Great, who wrote to Voltaire in 1778: "The English built ships on the most advantageous shape as indicated by Newton, and their admirals have assured me that these ships did not sail nearly so well as those built according to the rules of experience. . . . Vanity of vanities! Vanity of geometry!"[2]

Or, we concede that no mathematical model can perfectly predict the weather. A "perfect" weather forecasting equation would have to take into account such a blizzard of interacting variables—wind speed, barometric pressure, amount of sunlight, etc.—that the complexity would soon overwhelm the mathematics. This is not to say that we should give up. Weather forecasting continues to improve, and the mathematical models that describe it have grown increasingly sophisticated. But no model can accurately predict, for example, the *exact* number of raindrops that will fall on the roof of Dubuque's City Hall during the month of February. Such accuracy is utterly beyond us. Of course there *will* be a certain number of raindrops hitting that roof in February, and mathematicians' inadequacy will not stop it from raining. In the perceptive words of Augustin Fresnel, "Nature is not embarrassed by difficulties of analysis."[3]

In what follows we shall try to avoid embarrassment. Our goal is to select from the innumerable applications of mathematics two that are simple and yet reveal something significant about the world in which we live. The first applies mathematics to the measurement of space and the second to the measurement of time.

Consider this situation: We stand on one side of a river directly opposite a tall evergreen tree. Unfortunately we can't swim and are afraid of heights. Under these limitations, how can we find the size of the tree?

The answer lies in trigonometry, an old and extremely useful branch of mathematics. The name suggests its content: *tri* (three) *gon* (side) *metry* (measure)—the measurement of three-gons (i.e., triangles). More precisely, trigonometry exploits the similarity properties of right triangles.

Consider the right triangles in Figure U.1. Each contains an angle of 40° as well as a right angle, so the remaining angles in each are 180° − 90° − 40° = 50°. Because the triangles have angles respectively equal, they are similar and their corresponding sides are proportional. For example, the ratio of the side opposite the 40° angle to the

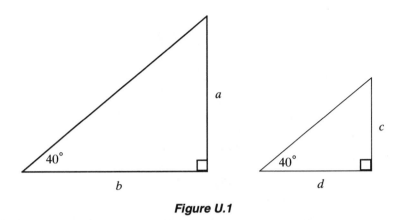

Figure U.1

side adjacent to the 40° angle is the same for the triangle on the left as for that on the right. Symbolically,

$$\frac{a}{b} = \frac{c}{d}$$

So if we know that $a = 83.91$, $b = 100$, and $c = 55$, we can substitute and cross-multiply to find the unknown side:

$$\frac{83.91}{100} = \frac{55}{d} \rightarrow 83.91d = 5,500 \rightarrow d = \frac{5,500}{83.91} = 65.55$$

Using proportionality and knowledge of three sides, that of the fourth can be determined.

This makes it appear that a *pair* of right triangles is necessary, but there is no reason both must be part of the stated problem. That is, if we were given just the right-hand triangle of Figure U.2, could we find d anyway?

The answer is "yes," for we can easily imagine another, albeit ideal, right triangle with a 40° angle and then determine the unknown ratio in a purely mathematical fashion. Adopting this viewpoint, trigonometers define the **tangent** of an angle α in a

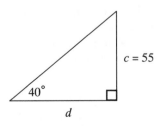

Figure U.2

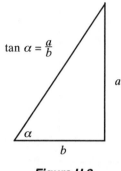

$$\tan \alpha = \frac{a}{b}$$

a

α

b

Figure U.3

right triangle—denoted by tan α—as the ratio of the leg opposite the angle to the leg adjacent to the angle. In Figure U.3,

$$\tan \alpha = \frac{\text{opposite leg}}{\text{adjacent leg}} = \frac{a}{b}$$

This ratio can be calculated without recourse to measurements of physical triangles. Greek mathematicians Hipparchus and Ptolemy did as much 2,000 years ago, and later work by Indian and Arabic scholars generated trigonometric tables giving tan α for whatever angle α is desired. These discoveries have found their way into the modern hand-calculator which, with a few taps on the keys, gives tan 40° = 0.8390996.

Returning to the lone triangle in Figure U.2, we then reason trigonometrically:

$$\tan 40° = \frac{\text{opposite}}{\text{adjacent}} \rightarrow 0.8390996 = \frac{55}{d}$$

Therefore $0.8390996d = 55$ and $d = 55/0.8390996 = 65.546$, the answer obtained above.

The key point is that mathematicians can compute the tangent ratio for an ideal right triangle and use this as one of a pair of similar triangles in solving real-world problems. That is precisely what we shall do to find the height of the tree on the opposite bank.

The first objective is to determine the river's width. This can be done by walking down the shore a measured distance—say, 100 feet—and then finding the angle from our new position to the evergreen. Suppose this is 58°. The configuration seen from above appears in Figure U.4. In right △ABC, the river's width is of unknown length *b*, side *BC* has length 100 (carefully measured along the riverbank), and ∠ABC has measure 58°. Thus,

$$\tan 58° = \frac{\text{opposite leg}}{\text{adjacent leg}} = \frac{b}{100}$$

Top view

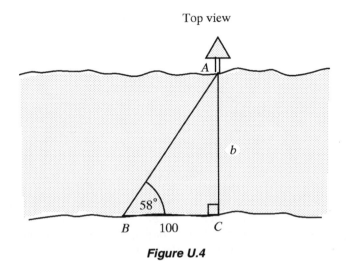

Figure U.4

Cross-multiplication gives $b = 100 \times \tan 58° = 100 \times 1.600 = 160.0$ feet, where the numerical value of $\tan 58° = 1.600$ was furnished by a calculator. This is the width of the river.

But we are not done, for the tree's height remains unknown. We simply walk back to the original point directly across the river from the evergreen and measure the angle to the tree's top. Suppose this comes out at 30°. Again we have generated a right triangle, this one perpendicular to the ground as shown in Figure U.5. The triangle's base is the river's width, previously shown to be 160 feet; the evergreen's height is the unknown x; and the angle is 30°. Again invoking the tangent ratio we find:

$$\tan 30° = \frac{\text{opposite leg}}{\text{adjacent leg}} = \frac{x}{160}$$

and so

$$x = 160 \times \tan 30° = 160 \times 0.57735 = 92.4 \text{ feet}$$

We have found the height of the evergreen without leaving the ground or getting our feet wet. Although this is a simple application, it suggests an undeniable power.

Of course, a devil's advocate could dismiss the previous work as unnecessary. After all, someone could get a canoe, row across the river, chop down the tree, and measure its height. It is not as though trigonometry yielded information that was *impossible* to obtain otherwise.

To blunt such indifference, we offer a more dramatic example of trigonometry in action. It takes us back to 1852 and to the office of the surveyor general of India, where trigonometric measurements from an ambitious British survey of the Himalayan Range were being used to calculate the heights of those distant peaks.

Side view

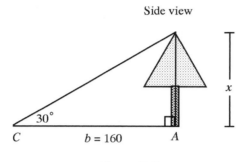

Figure U.5

Acquiring this information was enormously complicated. For one thing, there were problems of scale. Unlike a tree across the river, the Himalayas were a hundred miles from the surveyors. At that distance one must take into account atmospheric distortion and the curvature of the earth. Political quarrels had kept surveying teams from entering either Nepal or Tibet, the two nations along whose border runs the great Himalayan chain. Only because of their incredible size could the peaks be seen at all from the Indian foothills; a lesser range would have lain invisible over the horizon.

In spite of these difficulties, work proceeded. Readings were made of various peaks, and the data were analyzed by clerks at the Surveyor General's Office. It was there, according to mountaineering lore, that the Bengali Chief Computer (a human being, not a machine) checked and rechecked his calculations before announcing excitedly that he had discovered the highest mountain on earth.[4]

The survey had labeled it only as Peak XV. It did not, in fact, *look* like the highest of the mountains on the horizon, but that was because it was much further away (its distance, like the river's width in the example above, was also calculated from the trigonometric data). Peak XV rose more than 29,000 feet, or five and one-half miles above sea level. Its summit scraped the stratosphere. By comparison, Mt. Blanc, Europe's highest peak, stood roughly two and one-half *miles* lower.

The British, as is customary for colonial powers, named the peak after one of their own: Sir George Everest, former head of the trigonometric survey. Of course, the presence of the great mountain was no secret to those who inhabited its foothills. The Tibetans to the north had long called it Chomolungma, meaning "Goddess Mother of the World," and the Sherpas to the south had named it Sagarmatha, meaning "Mother of the Universe." Nonetheless, it is by the name Mt. Everest that the earth's highest peak is most widely known. If this is a vestige of British imperialism, at least we concede that the name "Everest" has a certain majesty befitting such a wonder of nature. Had the survey director been named Sir George Terwilliger, the effect would have been considerably diminished.

The point of this illustration, of course, is that the mountain's height was found trigonometrically in 1852. This was more than a century before human beings, in the persons of Tenzing Norgay and Edmund Hillary, first stood atop Everest's remote summit in late May of 1953. Climbing the mountain required backpacks and ice axes and extraordinary courage. Determining its altitude required only trigonometry.

If this earthbound example suggests the utility of mathematics, we shall provide an even more remarkable illustration. The same reasoning that measured evergreens and Everests was employed to measure the unimaginably greater distances to the moon, to the sun, and to the planets.

This story dates back at least to Greek and Islamic scholars, whose estimates of solar and lunar distances were based upon naked eye observations of eclipses and a knowledge of trigonometry. Around A.D. 850, for instance, the astronomer Abdul'l-Abbas Al-Farghani calculated the average distance of the sun from the earth to be about 1,170 times the earth's radius. This is a gross underestimation, for it puts less than 5 million miles between us and the sun, a distance that, if true, would fry our planet to a cinder. But it was a start.[5]

With the coming of the telescope in the seventeenth century more accurate observations were possible. These were essential in calculating triangles stretching from the earth to the sun, for which small inaccuracies in measurement resulted in errors not of feet, as with the evergreen, but of millions upon millions of miles. The required precision pushed the instruments of the day to their limits. In spite of such challenges, by the end of the seventeenth century Giovanni Cassini (1625–1712) calculated the distance of the sun at about 22,000 earth radii.[6] This translates to about 87 million miles (compared to the currently accepted figure of 93 million miles). It was a splendid example of the usefulness of trigonometry in resolving a seemingly impossible extraterrestrial problem.

As often happens in science, the solution of one problem makes possible the solution of another. In this case, knowledge of the solar distance led to the first estimate of the velocity of light, one of the most significant constants in all of physics. The following tells how it was done.

Early in 1610, Galileo had discovered, through his "spyglass," four moons circling the planet Jupiter. Astronomers subsequently recorded the movements of these distant satellites so that by the 1670s Cassini had prepared accurate tables giving the times at which the innermost moon, Io, would disappear behind the giant planet. These eclipses of Io should take place every 42 hours and 27 minutes.

But an unexpected phenomenon was observed. The disappearance of Io behind Jupiter occurred later than predicted when the earth and Jupiter stood on opposite sides of the sun from one another (shown on the left in Figure U.6) but earlier than predicted when the two planets were lined up on the same side of the sun (as on the right of Figure U.6). There seemed an inexplicable irregularity in the motion of Io about Jupiter.

Taking note of this delay was one of Cassini's assistants, Ole Roemer (1644–1710). He wondered what could account for the tardiness of the eclipse when the planets were farthest apart and the gradual quickening as they moved closer together. One explanation, of course, was that Io circled Jupiter at varying speeds, going faster when the earth was near but then slowing down as the earth receded.

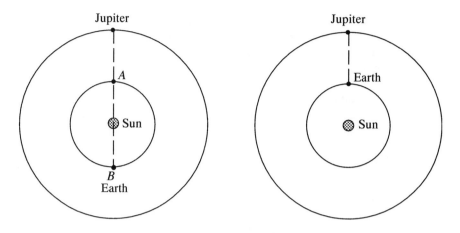

Figure U.6

Unfortunately this violated the laws of physics and, in any case, what would a moon of Jupiter care about the whereabouts of the earth?

The simpler explanation, and one that Roemer endorsed, was that Io moved at a uniform velocity but its light took longer to reach us when it had farther to travel. The apparent delay was caused not by anything happening in the vicinity of Jupiter but by the extra time consumed as light crossed our orbit to reach earthly eyes.

Of course, people knew that *sound* takes time to get from one point to another, as is readily observed in the delayed report of thunder after a faraway lightning bolt. But there was a widespread belief that *light* travels instantaneously, that what happens at one place will be seen immediately by anyone, anywhere in the universe. So thought such authorities as Aristotle in classical times and Descartes in the early seventeenth century. But Roemer explained the apparent slowing down and speeding up of Jupiter's moon precisely as one would explain the delayed sound of thunder, with the critical exception that it was now *light* that took time to get from place to place.

Roemer himself was less interested in determining the actual velocity of light than in proving that its transmission was not instantaneous.[7] But we can use Roemer's data to produce a "seventeenth-century" estimate of its speed. Observations indicated that eclipses of Io fell behind by 22 minutes as the earth moved from its closest to its farthest distance from Jupiter. Roemer attributed this lost 22 minutes to the time necessary for light to cross the diameter of the earth's orbit, that is, to move from point A to point B in Figure U.6. Thus, light takes half of this—that is, 11 minutes—to cover the distance from earth to sun. If we use for this distance Cassini's estimate of 87 million miles (obtained by trigonometry), we conclude that light travels about $87,000,000/11 = 7,910,000$ miles in one minute, or $7,910,000/60 = 132,000$ miles in a second.

This was an extraordinary speed. Christiaan Huygens exclaimed, "I have learned recently with much joy of the beautiful discovery made by M. Römer, to demonstrate that light in spreading from a source requires time, and even to measure this time. It is a very important discovery."[8] Another astronomer of the period observed in astonishment that "we shall be terrified both by the immensity of the distances, and by the rapidity of the movement of light."[9]

As it turned out, even this speed was too low. The radius of the earth's orbit had been underestimated by a good 6 million miles and the time required for light to cross it was overestimated by many minutes. In fact, it takes light a bit more than 16.5 minutes to make the crossing rather than the 22 minutes of Roemer's estimate. The velocity of light is now taken to be 186,282 miles per second.

So, mathematics certainly proved useful in measuring vast distances across space. But the chapter's other example is no less remarkable: the use of mathematics to measure vast distances back through time.

For centuries scholars had assigned *relative* ages to prehistoric objects by the simple observation that, as we dig down through layers of soil and rock, we move backward into the past. That much is easy. But what of the *absolute* ages of an unearthed antler, an Egyptian burial shroud, a charred piece of wood from a cave dwelling? How can an archaeologist hope to determine the decade, the century, or even the millennium from which these items came? Such information seems unknowable and forever lost.

But it is not. One of the most impressive attributes of science is its will to know, even in the face of situations that seem hopeless. In the charming words of Sir Thomas Browne, "What song the Sirens sang, or what name Achilles assumed when he hid himself among women, though puzzling questions, are not beyond all conjecture."[10]

It was in that spirit that chemist Willard Libby and his associates made the great discovery of radiocarbon dating in the years following World War II. For this work Libby received chemistry's Nobel Prize in 1960, a well deserved recognition for having unraveled the mystery of ancient campfires or prehistoric skeletons. What Libby discovered was that those old bones and shards of wood are actually tiny, accurate clocks. And to decipher their hidden message required a knowledge of the chemical properties of carbon and of the mathematical properties of the natural logarithm.

First the chemistry. Carbon comes in three varieties. Two of these, the more plentiful and stable, are called carbon-12 and carbon-13; the other, rarer and more transient, is carbon-14, a radioactive isotope with a half-life of about 5,568 years. *Half-life* is a technical term with a simple meaning: In 5,568 years, half of an original quantity of carbon-14 will be lost to radioactive decay. Thus, a pound of carbon-14 today, if left undisturbed, will have decayed to just half a pound 5,568 years from now and to just a quarter of a pound 5,568 years after that.

Carbon-14 originates from cosmic radiation in the upper atmosphere where it reacts with oxygen to form radioactive carbon dioxide. This eventually settles to the earth's surface as part of the carbon stew in which all life exists. Libby put it quite directly: "Since plants live off the carbon dioxide, all plants will be radioactive; since the animals on earth live off the plants, all animals will be radioactive."[11] As a consequence, radioactive carbon is present in the carrots you had for lunch, in the petunias from your garden, in your pet hamster, and in the vice-president. It is a shared mark of our earthly origins.

With some sophisticated chemistry, it is possible to determine the proportions of radioactive and nonradioactive carbon in living tissue, and it is reasonable to assume that similar levels existed in the animals and plants of the past. As organisms go about the business of living, they continually replenished lost carbon-14 from the food chain and thereby maintain a fairly constant equilibrium in these proportions.

But at the instant a mastodon dies or a tree is felled, its carbon-replenishing days are over. Whatever carbon existed in its tissues is all that will ever be there. As the ages pass, the nonradioactive carbon remains unchanged but the carbon-14 undergoes radioactive decay—which is to say, it tends to disappear. The relative proportion of radioactive to nonradioactive carbon therefore diminishes over time. Like an old clock winding down, the radioactive emissions become proportionately slower. This decline in carbon-14 begins upon the death of the organism and continues right down to the day the old bone or wooden trinket is dug out of the earth.

Chemists, using specialized equipment, can determine the current radioactive output of an artifact's carbon-14—a lowered output from what it was when alive. Because we know the rate at which carbon-14 decays, we can calculate within certain bounds of precision how long it has taken the item to reach this diminished radioactivity level. This, of course, is precisely the length of time since the bone or wood ceased to be part of a living creature; to put it more succinctly, it is the object's *age*. We have here an amazing piece of scientific detective work, one indeed worthy of a Nobel medallion.

But as is so often the case in the sciences, tidying up the final details requires mathematics. For radiocarbon dating, the key equation is

$$A_s = \frac{A_o}{e^{0.693t/5568}}$$

where A_s is the current radioactivity level of the relic, A_o is the radioactivity level of a living item of the same type, and t is the time that has passed since its death. Note that, embedded within this equation, is the carbon-14 half-life of 5,568 years. And note also the presence of the number e in a starring role.

The following example, similar to one Libby himself considered, illustrates the mathematics involved.[12] Suppose archaeologists have unearthed a fragment of wood from a funerary ship of a long-gone Egyptian pharaoh. We assume that the tree from which the wood came was cut more or less at the time of the pharaoh's death. Chemists analyzing the wood in the lab determine its current radioactivity level to be $A_s = 9.7$ decompositions per minute per gram of carbon. By contrast, freshly cut pieces of the same kind of wood produce $A_o = 15.3$ radioactive decompositions per minute per gram of carbon. The object is to calculate t, the age of the wood.

Substituting A_s and A_o into the equation yields:

$$9.7 = \frac{15.3}{e^{0.693t/5568}}$$

Cross-multiply to get $9.7e^{0.693t/5568} = 15.3$, and so $e^{0.693t/5568} = 15.3/9.7 = 1.577$.

Now, the objective is to determine the unknown t in the exponent. We first take the natural logarithm of both sides of the equation:

$$\ln(e^{0.693t/5568}) = \ln 1.577$$

A quick reference to Chapter N reminds us that $x = \ln(e^x)$. Thus we conclude

$$\frac{0.693t}{5568} = 0.456$$

where the value of ln 1.577 is furnished by a calculator. It follows that

$$0.693t = 5{,}568 \times 0.456 = 2{,}539.0 \rightarrow t = \frac{2{,}539.0}{0.693} = 3{,}663.8 \text{ years}$$

Our calculations therefore reveal that the funerary boat was constructed, and the pharaoh died, 3,664 years ago. Of course, the precision of this estimate is subject to doubt; everything from inaccurate determination of radioactivity levels to contamination of the sample could affect our result in some fashion. Yet if we asserted the pharaoh died about 3,700 years ago, we would be on pretty solid ground. Combining knowledge of both wooden logs and mathematical ones, we have forced a mute object to yield the secret of its own antiquity. Thanks to chemistry and mathematics, the door to the past has swung open.

Whether measuring the height of Everest, the speed of light, or the antiquity of a pharaoh, mathematics has proved its usefulness beyond a shadow of a shadow of a doubt. Morris Kline went so far as to assert that "the primary value of mathematics is not so much what the subject itself offers but what it helps man to achieve in the study of the physical world."[13]

Many would argue that in this passage Kline went overboard. He seems to be suggesting that, if the astronomers and chemists suddenly satisfied all of their mathematical needs, then mathematicians could simply clean out their desks and retire.

An opposing view comes from G. H. Hardy, among the purest of the pure mathematicians. Hardy, with his own knack for outrageous statements, conceded that "a good deal of elementary mathematics . . . has considerable practical utility" but went on to contend that useful ideas are, "on the whole, rather dull; they are just the parts which have least aesthetic value. The 'real' mathematics of the 'real' mathematicians, the mathematics of Fermat and Euler and Gauss and Abel and Riemann, is almost wholly 'useless.' "[14]

Although most mathematicians would retreat a bit from Hardy's unflinching embrace of uselessness, there is widespread agreement in the profession that mathematics is not merely a servant of the sciences. Results such as the prime number theorem from Chapter P retain a beauty and fascination—and hence a mathematical legitimacy—in spite of having no practical value at all. When we judge mathematics solely on utilitarian grounds, we ignore one of the central privileges of being human: the opportunity to soar intellectually for the sheer joy of soaring.

Although the truth probably lies somewhere between the positions of Morris Kline and G. H. Hardy, the utility of mathematics is inescapable, and mathematicians by the tens of thousands continue to direct their efforts toward its applications. Among mathematicians one sometimes hears the following pearl of wisdom: It is easy to become a mediocre applied mathematician; somewhat more demanding is to become a mediocre pure mathematician; considerably harder is to become an outstanding pure mathematician; but most difficult of all is to become an outstanding applied mathematician. To excel in the application of mathematics one must master multiple subjects: mathematics as well as astronomy or chemistry or engineering. And, whereas a pure mathematician can freely modify postulates or add hypotheses to make the work easier, the applied mathematician must make do with the uncontrollable phenomena of the external world. Pure mathematics is driven by logic; applied mathematics is driven by logic *and* nature. Pure mathematicians can change the ground rules; applied mathematicians are stuck with what reality gives them.

We conclude with the words of Galileo, a scientist of the very first rank who heard mathematical echoes reverberating from every corner of the natural world. The utility of mathematics was never more succinctly addressed than in Galileo's description of the universe as a "grand book" that "cannot be understood unless one first learns to comprehend the language and read the letters in which it is composed. It is written in the language of mathematics."[15]

Venn Diagram

All A is B

In the mid-nineteenth century, John Venn (1834–1923), a Fellow of Cambridge University, devised a scheme for visualizing logical relationships. Venn was a cleric in the Anglican Church, an authority on what was then called "moral science," and the compiler of a massive index of all Cambridge alumni. He was not terribly outstanding in mathematics. But, for better or worse, a single contribution has made him immortal.

That contribution is the Venn diagram. It is as much a fixture of today's textbooks as the title page or table of contents. A *Venn diagram* is simply a field within which circular areas represent groups of items sharing common properties.

For instance, within the universe of all animals (the large rectangle in Figure V.1), region C represents the camels, region B the birds, and region A the albatrosses. A glance at the diagram reveals that

- All albatrosses are birds (region A lies entirely within region B).
- No camels are birds (regions C and B are nonintersecting).
- No camels are albatrosses (regions C and A are nonintersecting).

This is a depiction of a basic rule of logic—namely, that from the statements "all A is B" and "no C is B," it follows that "no C is A." The conclusion is evident when we look at the diagram's circles.

No one, not even John Venn's best friend, would argue that his underlying idea is very deep. Venn's innovation took immeasurably less brainpower than, for instance, Archimedes' determination of spherical surface from Chapter S. The latter required

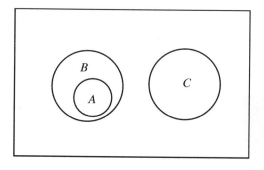

Figure V.1

extraordinary insight; the former might just as well have been discovered by a child with a crayon.

But there is more. Gottfried Wilhelm Leibniz, often regarded as the founder of symbolic logic, used little diagrams of this sort in the seventeenth century. And in Leonhard Euler's *Opera Omnia* we find the illustration depicted in Figure V.2. Look familiar? This is a "Venn diagram" a century before Venn. If justice is to be served, we should call this an "Euler diagram." Of course, such a name change would add little to Euler's stupendous fame but would obliterate John Venn's reputation altogether.

So, the Venn diagram is neither profound nor original. It is merely famous. Somehow within the realm of mathematics, John Venn's has become a household name. No one in the long history of mathematics ever became better known for less. There is really nothing more to be said.

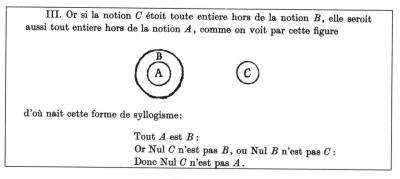

Figure V.2

A "Euler diagram"

(Courtesy of Lehigh University Library)

Where Are the Women?

If the reader is keeping score, it should be apparent that more men than women have appeared on the pages of this book. This imbalance reflects the historical male dominance of the mathematical sciences. But does it mean women have never contributed to the subject, that women are not currently contributing, or that women will not contribute in the future?

The answers to these questions are, respectively, "no," "of course not," and "get serious." Women appear in the history of mathematics as far back as classical times and are far more active today than ever before. Their presence comes in spite of obstacles that male mathematicians can hardly imagine, among which are not only a lack of encouragement but an active *discouragement* of women's participation.

At the outset we acknowledge that a short list of history's most influential mathematicians—Archimedes, Newton, Euler, Gauss—is exclusively male. Women in mathematics prior to 1900 are relatively few in number. Among those most often cited are Hypatia of Alexandria, who lived around A.D. 400. Emilie du Chatelet (1706–1749) and Maria Agnesi (1718–1799) were active during the eighteenth century, and Sophie Germain (1776–1831), Mary Somerville (1780–1872), and Ada Lovelace (1815–1852) worked in the early nineteenth. Sofia Kovalevskaia (also known in the literature as Sonya Kovalevsky) brings the list up to the threshold of the twentieth century.

Of these, Hypatia was an influential geometer, teacher, and writer; Chatelet was known for translating Newton into French and Somerville for translating Laplace into English. Agnesi published a mathematics text in 1748 for which she received well-deserved recognition. Lovelace worked with Charles Babbage in the production of his "analytical engine"—a proto-computer.

Germain and Kovalevskaia are the most accomplished mathematicians on the list. The former worked in both pure and applied mathematics. We mentioned her work on Fermat's last theorem in Chapter F, and in 1816 she received a prize from the French Academy of Science for her mathematical analysis of elasticity. Kovalevskaia completed a doctorate and held a university position, ground-breaking achievements for a woman of her day. In the process she earned the respect of many otherwise skeptical male colleagues.

So, women mathematicians certainly existed prior to the twentieth century. What is surprising is not that there were few but that there were any at all. For women had to overcome not only the normal barriers facing those with mathematical aspirations—namely, that higher mathematics is extremely hard—but an assortment of culturally imposed barriers as well. We shall discuss three of the most serious impediments blocking their paths.

First is the pervasive *negative attitude* about women in the discipline, an attitude that has left its mark upon both males and females. At the core is a belief that women are incapable of doing serious mathematics. This conviction has been held by large numbers of people, including some very influential ones. Immanuel Kant is said to have remarked that women might as well have beards as "worry their pretty heads about geometry," a most discouraging remark from so important a philosopher.[1] Unfortunately, such attitudes are by no means a thing of the past. There are numerous accounts of contemporary high school girls wishing to sign up for trigonometry or calculus only to be advised by counselors, parents, or friends that courses in home economics or English are better suited to the feminine way of thinking. Believe it or not, this sort of thing is still going on.

Among the arguments advanced to prove that women cannot do mathematics is that so few have done it. In other words, the absence of women in mathematics is used to confirm their unfitness for the subject. Of course, reasoning along these lines is ridiculous. It is not unlike attributing the absence of African Americans in major league baseball prior to World War II as proof of some fundamental inability to play the game. As Jackie Robinson, Henry Aaron, and so many others have amply demonstrated, the dearth of black major-leaguers reflected not a lack of ability but a lack of opportunity.

That women can do mathematics is evident from the individuals mentioned above. We could augment this with an extensive list of women who have been active in more recent times, from Grace Chisholm Young who played a key role in refining the advanced theory of integration during the early years of the twentieth century; to Julia Robinson, solver of Hilbert's tenth problem; to Emmy Noether, one of the twentieth century's most accomplished algebraists. Attitudes that women cannot do mathematics are groundless.

But there has also been a simultaneous attitude that women *should not* do mathematics. At best, it was deemed a waste of time; at worst, it was seen as harmful.

Just as children should not go near the freeway, so women should not go near mathematics.

As an example we cite Florence Nightingale, who later made her name in the medical arts. In her youth she displayed a mathematical enthusiasm that prompted her incredulous mother to ask, "what use are mathematics to a married woman?"[2] As we noted in Chapter U, few human enterprises are more useful than mathematics, yet Nightingale was told just the opposite. Given the traditional roles available to a nineteenth-century woman, mathematics was seen as having no utility whatever.

Then, too, women were advised that studying mathematics would destroy their social graces. Worse, medical evidence purportedly showed that women who thought too much would experience a diversion of blood from the reproductive organs to the brain, with obvious dire consequences. It is interesting that men (of all people) never seemed to worry about an analogous blood flow.

Attitudes of this sort quickly translated themselves into action, or perhaps more accurately, into inaction. Germain had to write her mathematics under a male alias; Kovalevskaia was at first denied an academic position in spite of her proven abilities. Even the great Emmy Noether encountered hostility when she sought a junior position at Göttingen University in Germany. Her detractors disapproved, fearing that once a woman got her foot in the door, there would be no stopping the downhill slide. To this, David Hilbert responded with a delicious bit of sarcasm: "I do not see that the sex of the candidate is an argument against her admission. After all, we are a university and not a bathing establishment."[3] Noether eventually got her job, and the mathematical community survived quite nicely.

A second impediment has been the *denial of formal education*. Mathematics is a subject that requires training—intense, extensive training. To reach the frontiers, one must march through the preliminaries, and for a discipline as old and complex as mathematics, this takes years of effort. In the past it was rare for a woman even to *begin* this arduous route. Consequently, success in higher mathematics became almost impossible.

How did men learn the subject? Often they were tutored or received one-on-one instruction. We have observed that Leibniz sought out Christiaan Huygens and that Euler studied with Johann Bernoulli. Here was an established master passing the torch to a future one. Very few women had similar opportunities.

Males, after suitable training, would head off to university, where their talents and abilities could be further nurtured. Gauss attended the University of Helmstadt, Wantzel attended the École Polytechnique, and Russell attended Cambridge.

Contrast this with Germain, an individual of enormous promise who, because of her gender, was prohibited from entering the university lecture hall. Only by listening at the classroom door or copying notes from sympathetic male colleagues could she clandestinely keep up with the material. That she prevailed is a testament, in Gauss's words, to a woman of "the most noble courage."[4]

The overwhelming majority of women thus had no real contact with the world of higher mathematics. It is worth noting that many of the women mentioned above were well-to-do, with the associated advantages of privilege. Germain had access to her father's library. Somerville eavesdropped on her brother's tutoring sessions. These daughters of affluent families clearly had options not available to those of more modest means. As Michael Deakin observed about *poor* women with mathematical promise, "the twin disabilities of poverty and womanhood were clearly too much."[5]

It is interesting to compare this situation with that of women writers from roughly the same period. Reading and writing were part of a gentlewoman's training, although these were regarded as necessary social skills rather than the means toward an artistic career. Nonetheless, many a woman acquired the writer's tools. Then, if she had sufficient time, discipline, and ability, she could apply these tools toward the creation of poetry or literature. An example is Jane Austen, whose writing grew out of the life she saw around her, carefully observed and filtered through her extraordinary talent. Austen could read, she could write, and she was an artist. She produced a body of work that places her among the great names of English literature.

It is true that many girls also learned a bit of elementary ciphering. But the parallels with literature break down at this point. Progress in higher mathematics requires an understanding of geometry, calculus, differential equations—each building upon its predecessor. Very few people acquire such knowledge without training. When women were denied that training, they were simultaneously denied the tools of the mathematician. Doors slammed shut on their scientific futures. We can never know the mathematical Jane Austens who were lost for want of formal education.

That was the past. What about the present? Certainly the explicit barriers have fallen, and universities no longer impose a prohibition on women of the sort that Germain encountered. On the contrary, there is cause for optimism in the most recent data on mathematics enrollments in U.S. colleges. During the 1990–91 academic year, U.S. institutions granted 14,661 undergraduate degrees in mathematics. Of these 6,917, or 47 percent, went to women. This near perfect split would have been unimaginable to the male mathematical establishment of a century ago.

When we look at advanced degrees, the data are less encouraging. In the same academic year, women received two-fifths of the master's degrees but only one-fifth of the doctorates in the mathematical sciences.[6] This suggests that women, although pursuing undergraduate mathematics in impressive numbers, are much less likely to continue their training into graduate programs. And since it is from here that tomorrow's research mathematicians and college professors will come, the situation remains imbalanced.

Why do women not continue to graduate school? Historically, many women have aspired to be teachers at the pre-university level and thus had no need for a research degree. In some cases low self-esteem, fostered by attitudes described above, surely has contributed to a pessimism about grad school success. And there is the matter of encouragement, of finding a mentor who can buck up the spirits and smooth out the difficulties that come with the study of higher mathematics. Men have an abundance of compatriots and role models; women can feel alone on the stormy academic seas. Their route to formal education remains in many ways different from that of their male counterparts.

Even when women overcame negative attitudes and acquired a solid education, there was yet another obstacle: the *lack of support* for them to pursue their work apart from the demands of everyday life.

Doing mathematics requires large blocks of otherwise unencumbered time. Research mathematicians spend long periods just sitting around thinking. In the past, as today, blocks of time are not equally available to all. As suggested above, the simplest expedient is to be independently wealthy. Legend says that Archimedes was part of the royal family of Syracuse. The marquis de l'Hospital (1661–1704) was rich enough to hire Johann Bernoulli to instruct him in the new calculus that was then sweeping Europe. Of the women on our list, Emilie Chatelet was a marquise, Lovelace was a countess, and Agnesi was the child of wealth. None of these folks spent their days doing laundry.

Another source of support was the European Academy, the think tank of its day. Patronage from academies in Berlin, Paris, or St. Petersburg put food on the tables of any number of scholars. Euler, who held positions at both Berlin and St. Petersburg, was a mathematician who used these opportunities to the fullest.

Or one could have a job so undemanding as to allow periods of leisure for study and contemplation. We have noted that Leibniz somehow found time during his diplomatic mission in Paris to learn mathematics and eventually develop calculus. The magistrate Fermat seemed never to have enough to do in court, so he did mathematics instead.

In summary, it did not hurt for the potential mathematician to be a person of means, a member of an academy, or somewhat underemployed. Today, of course, the primary support for mathematicians is the research university, which provides an office, a library, money for travel, like-minded colleagues, and modest teaching responsibilities. In return, the mathematician is expected to think, and think deeply, about the frontiers of the subject.

Contrast this with the historical role of women: to stay at home, to raise children, to cook and sew and attend to the domestic chores while husbands or brothers work outside the home. Even if she had the training, where did a woman get the time to think about differential equations or projective geometry? The expectations upon her were entirely different.

In fact, a woman seldom even had a *room* of her own. As Virginia Woolf reminds us in an essay of that title, women rarely had a special place to go to be alone, to think, to write (or to do mathematics). Woolf spins a fable of Shakespeare's imaginary sister Judith, every bit as talented as her brother, whose life was devoted to serving the daily needs of the family as William perfected the writer's craft. According to Woolf, Shakespeare's sister

> was as adventurous, as imaginative, as agog to see the world as he was. But she was not sent to school. She had no chance of learning grammar and logic, let alone of reading Horace and Virgil. She picked up a book now and then . . . and read a few pages.

But then her parents came in and told her to mend the stockings or mind the stew and not moon about with books and papers.[7]

One sibling was the provider of support; the other was the recipient. It is a significant dichotomy.

Consider Leonhard Euler, the father of 13 children. Someone had to feed them, change their diapers, and clean their clothes. It was not Leonhard. Consider Srinivasa Ramanujan (1887–1920), an incredibly gifted mathematician from early in the twentieth century. In his day-to-day affairs, he exhibited an almost childlike helplessness, and his wife cared for his every need. Consider Paul Erdös, whom we encountered in our first chapter as he learned how to butter his bread at the age of 21. Obviously, he received extraordinary support from his mother through those early years of mathematical discovery.

What if the shoe had been on the other foot? Would Mrs. Euler or Mrs. Ramanujan or Mrs. Erdös have succeeded in mathematics had her daily needs been met by another? Would these women have been famous had they had blocks of time to devote to mathematical contemplation? No one will ever know. But would more women appear in the annals of mathematics if they had received the same kind of support as these men? Undoubtedly.

All of the impediments cited above—negative attitudes, difficulty in getting a mathematical education, lack of a support system—were present in the life of Sofia Kovalevskaia, "the greatest woman mathematician prior to the twentieth century."[8] Her story illustrates the problems and the triumphs of women in mathematics.

Kovalevskaia was born in Moscow early in 1850 and grew up in an affluent household with a scholarly atmosphere, an English governess, and the chance to learn mathematics. There is a charming story that the walls of her bedroom were papered with old lecture notes from her father's courses on calculus. The young girl was fascinated by these strange formulas that surrounded her like silent friends. She vowed someday to learn what secrets they held.

This, of course, took training. At first, she learned arithmetic. Then Kovalevskaia was permitted to attend her cousin's tutoring sessions, primarily as a way of shaming him into working harder. In this way she acquired an understanding of algebra (even if he did not). Next, Kovalevskaia borrowed a book written by a physicist who lived nearby. Upon reading it, she encountered difficulty with the trigonometry, a subject of which she was totally ignorant. Unwilling to give up but unable to find a suitable explanation, Kovalevskaia simply developed the ideas from scratch. Her physicist neighbor, when he realized what she had done, observed in wonder, "she had created that whole branch of science—trigonometry—a second time."[9]

This sort of achievement indicates supreme mathematical creativity. At the age of 17, she and her family traveled to St. Petersburg, where Kovalevskaia overcame her father's objections and received tutoring in calculus. Her talents were such that, had she been of the opposite sex, she would have immediately gone off to university. Unfortunately, this was not an option for a nineteenth-century Russian woman.

Her response to this disappointment was, from a modern perspective, an extreme one. At the age of 18 she arranged a marriage of convenience with a young scholar bound for Germany, where she hoped for greater educational opportunities. The man was Vladimir Kovalevskii, a paleontologist willing to engage in this "fictitious marriage" for its potential benefit to the liberation of women. The two of them set off for Heidelberg University to pursue their separate interests while maintaining the façade of matrimony.

Kovalevskaia excelled at Heidelberg as she had elsewhere, so in 1871 she set her sights even higher: to the University of Berlin and its revered senior mathematician, Karl Weierstrass (1815–1897). The determined Kovalevskaia arranged a meeting with the world famous scholar to plead for private tutoring. Weierstrass sent her away with a set of problems so challenging that he expected never to see her again.

But he did. A week later, Kovalevskaia was back, solutions in hand. In Weierstrass's opinion, her work exhibited "the gift of intuitive genius to a degree . . . seldom found among even . . . older and more developed students."[10] She had added another skeptic—this time one of the world's most influential mathematicians—to her band of admirers.

With this began a long collaboration between the aging Weierstrass and the youthful Kovalevskaia. Her energy and insight earned his sincere respect, and he in turn put her in touch with much of the European mathematical community. Under his direction, Kovalevskaia did research on partial differential equations, Abelian integrals, and the dynamics of the rings of Saturn. This work earned for her a doctorate in mathematics from Göttingen in 1874. She was the first woman to receive such a degree from a modern university.

Throughout her life, Kovalevskaia was drawn not only to mathematics but to issues of social and political justice. A champion of liberal causes, she supported rights for women and freedom for the Poles. At one point she wrote for a radical newspaper. With her husband, she surreptitiously entered Paris during the Commune of 1871, when the city was surrounded by Bismarck's army. On this adventure Kovalevskaia was actually shot at by German soldiers. Once in Paris, she tended to the sick and wounded and made contact with the radical leaders of the besieged city. Here was an individual willing to act upon her social convictions.

But there is more. Besides being a scientist and revolutionary, she was a writer. Kovalevskaia authored novels, poems, and dramas, as well as *Recollections of Childhood*, an autobiographical account of her formative years. While a Russian adolescent, she had met Fedor Dostoevskii and later in life came to know Ivan Turgenev, Anton Chekov, and George Eliot. This socially conscious mathematician moved in distinguished literary circles.

In short, Sofia Kovalevskaia possessed a striking diversity of talent. Brilliant, determined, articulate, she was described by a contemporary as "simply dazzling."[11] There emerges the picture of a riveting, charismatic human being, the sort about whom best-selling books or television miniseries are written.

As in all miniseries, along with triumph came tragedy. In spite of the peculiar circumstances of her marriage, she developed a genuine love for her husband, and the couple had a daughter in 1878. But five years later, after he had lost huge amounts of

Sofia Kovalevskaia on Soviet postage

money in failed business schemes, a despondent Vladimir Kovalevskii took his own life by ingesting chloroform. Sofia was now a widow and single parent.

Fortunately she was also a world-class mathematician. With the enthusiastic support of Gösta Mittag-Leffler, another of Weierstrass's disciples, she was appointed to the faculty of Sweden's Stockholm University. In 1889 her position became a lifetime professorship, another first for a woman in mathematics.

Her time at Stockholm was not without difficulties. The usual prejudice against women was compounded by Kovalevskaia's unflinching public support of progressive causes. Conservative scholars, unable to criticize her mathematics, condemned her contacts with a well-known German socialist. Both Weierstrass and Mittag-Leffler politely suggested that Kovalevskaia adopt a more discreet political posture. She did not.

On the mathematical side, she was named an editor of the journal *Acta Mathematica*, the first woman ever to hold such a post. She corresponded with mathematicians like Hermite and Chebyshev (whom we met in earlier chapters) and became an important link between the Russian mathematical community and its counterparts in Western Europe. In 1888 Kovalevskaia received the *Prix Bordin* from the French Academy of Science for her paper "On the Rotation of a Solid Body about a Fixed Point." With this came international fame, newspaper articles, and letters of congratulation. The acclaim was sufficient to win her a corresponding membership in Russia's Imperial Academy of Sciences (but still insufficient to earn her, as a female, an academic position in her homeland).

So, in 1891 a future of great promise seemed to lie before this remarkable individual. And then, unexpectedly, final tragedy struck. While on a trip to France, Kovalevskaia caught what seemed like a trivial cold. But as she returned to Stockholm, her condition worsened in the raw and wintry weather. Back home, she grew too weak to attend to her responsibilities. After falling into a coma, Sofia Kovalevskaia died on February 10, 1891, at the much-too-young age of 41.

As always when the talented die prematurely, she left behind shock, utter disbelief, and a host of unfulfilled dreams. Expressions of tribute arrived from across Europe and the attendant grief was genuine and widespread. We cannot know what additional contributions Kovalevskaia might have made to mathematics. Neither can we know what impact such contributions might have had upon the status of women in the discipline.

Whereas a talent like Kovalevskaia's is rare, women in mathematics have become increasingly common in the century since her death. And this, in turn, raises a troubling question. By devoting this chapter to women mathematicians, are we guilty of marginalizing women, of treating them as a kind of separate species? As women have moved into the professions of medicine and law, there is much less talk of "women doctors" or "women lawyers." In this chapter we do not want to suggest that the mathematical profession should be broken into two groups: mathematicians and women mathematicians. This certainly has not been our intent, and it is not the reality of the situation. Yet it is a danger.

Such was the opinion of Julia Robinson. As her prestige grew, as she entered the National Academy of Sciences and received her MacArthur Award, Robinson was cited as a triumphant woman in a male domain. "All this attention," she wrote in a very significant passage, "has been gratifying but also embarrassing. What I really am is a mathematician. Rather than being remembered as the first woman this or that, I would prefer to be remembered as a mathematician should, simply for the theorems I have proved and the problems I have solved."[12] The proper response is, "Amen!"

Although the inequities faced by women have yet to be eradicated, there is reason for optimism that Robinson's wish may be fulfilled. Mathematics enrollments have risen even as many of the prejudicial barriers have fallen. If the problem has not been fully solved, the evidence of progress is undeniable. It is hoped that, in the not too distant future, a chapter raising the question "Where are the Women?" will be seen as utterly unnecessary.

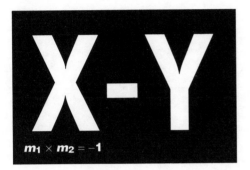

Plane

$$m_1 \times m_2 = -1$$

While consuming two letters of the alphabet at once, this chapter addresses a topic that has appeared repeatedly on the preceding pages, one so fundamental that it might seem to have existed forever.

We refer to the system of coordinate axes, the horizontal and vertical grids superimposed upon the plane that give numerical addresses to each point on the two-dimensional surface. The horizontal axis, the so-called x-axis, comes with a number scale increasing toward the right, and the vertical axis, the y-axis, has its scale increasing toward the top. With these, it is possible to go back and forth between a geometric point and its numerical coordinates.

Of course, plotting a single point is of little interest. The plot thickens when we have an equation—such as $y = x^2 + 1$ —and interpret it as the collection of *all* points (x, y) in the plane for which the relationship $y = x^2 + 1$ links the variables. Upon locating many such points, the algebraic equation generates a geometrical curve, in this case, the parabola in Figure XY.1.

This linkage between algebra and geometry seems quite natural. It is thus surprising to realize how recently it came into being. Whereas Euclidean geometry, sans algebra, dates back more than 2,000 years, this *analytic geometry* is not yet four centuries old. That makes it younger than, for instance, logarithms, *Romeo and Juliet*, and Boston.

The subject appeared, as did so many mathematical innovations, in the seventeenth century. The innovators were Pierre de Fermat and René Descartes, both French, both brilliant, and both important characters in the development of mathematics. As noted in an earlier chapter, Fermat's invention of coordinate geometry has been overshadowed by his better known contributions to number theory. Moreover,

273

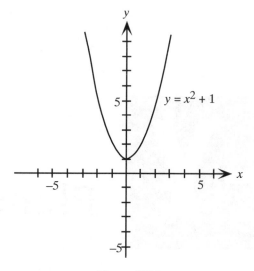

Figure XY.1

Fermat's delay in publishing diminished his influence, and by the time his work became available the novelty of the idea was long gone. Instead, the glory of analytic geometry goes to its first published proponent, René Descartes.

The year was 1637. Descartes had written a massive work called *Discours de la Methode*, a sort of philosophical road map for the scientific revolution. To this treatise he attached, as if an afterthought, an appendix titled *Géométrie*. Descartes began with this manifesto: "Any problem in geometry can easily be reduced to such terms that a knowledge of the lengths of certain straight lines is sufficient for its construction. . . . And I shall not hesitate to introduce these arithmetical terms into geometry."[1] The heretofore blank Euclidean plane, upon which idealized shapes played out their geometric roles, would now be invaded by numbers—Descartes's "arithmetical terms"—measuring their lengths and indicating their positions.

Unfortunately, most readers did not find *Géométrie* easy going. Even Isaac Newton confessed that at first he could make no sense of Descartes's method. Years later a biographer wrote that Newton

> [t]ook Descartes's Geometry in hand, tho he had been told it would be very difficult, read some ten pages in it, then stopt, began again, went a little farther than the first time, stopt again, went back again to the beginning, read on till by degrees he made himself master of the whole.[2]

If Newton had difficulty, imagine the plight of the less gifted student! Typical of Descartes was his admonition to the reader: "I shall not stop to explain this in more detail, because I should deprive you of the pleasure of mastering it yourself. . . . I find nothing here so difficult that it cannot be worked out by anyone at all familiar with ordinary geometry and with algebra."[3]

René Descartes
(Courtesy of Muhlenberg College Library)

Descartes was particularly blunt when he described his book to Mersenne. "I have omitted a number of things," he wrote, "that might have made it clearer, but I did this intentionally, and would not have it otherwise."[4] The implied philosophy—that clarity is to be avoided in mathematical exposition—is not recommended for the aspiring textbook writer.

Fortunately, others were able to recast these ideas in more intelligible terms. An edition of *Géométrie* prepared by Frans van Schooten (1615–1660) of Amsterdam hit the streets a dozen years after the Descartes original and, with the addition of a vast and helpful commentary, made the subject accessible to a much wider audience. It is significant that both Isaac Newton and Gottfried Wilhelm Leibniz, as they raced from relative ignorance to the discovery of calculus, benefited greatly from Schooten's edition of Descartes.

The subject they studied was not identical to its modern counterpart. In those days, the axes were not always drawn perpendicular to one another; sometimes the *y*-axis was not drawn at all; and the aversion to negatives often confined work to the upper right-hand region of the plane, the so-called first quadrant where both the *x* and *y* coordinates are positive. It would take awhile for all of this to get sorted out.

One who made a significant contribution was Newton himself, whose impact upon the subject tends to be lost in the glare of his other achievements. His *Enumeratio linearum tertii ordinis*, written in 1676 and finally published (after a typical Newtonian delay) in 1704, has been described as the work in which "analytic geometry may be said to have come into its own."[5] In it, Newton introduced, analyzed, and meticulously graphed 72 different kinds of third-degree equations. Obviously his enthusiasm for analytic geometry was exceeded only by his endurance.

And so, with the innovations of Descartes and Fermat and the subsequent contributions of Newton, the subject was established and standardized. It is easy for us to minimize the achievement, to write it off as a simple, obvious step. But history shows that what seems obvious *after* the fact may have been far from self-evident beforehand. As Julia Robinson wrote about a troubling mathematical problem:

> I have been told that some people think that I was blind not to see the solution myself when I was so close to it. On the other hand, no one else saw it either. There are lots of things, just lying on the beach as it were, that we don't see until someone else picks one of them up. Then we all see that one.[6]

These words perfectly describe the merger of geometry and algebra that occurred in the seventeenth century.

From the outset it was evident that analytic geometry has two important but opposite themes. In one, algebra is employed in the service of geometry; in the other geometry is used in the service of algebra. Viewed jointly, these produce a kind of mathematical symbiosis in which each facet of the problem benefits from its relationship to the other.

In large part Descartes was an advocate of the former. That is, he tended to begin with a geometric problem and apply algebraic techniques to reach a solution. For him the relatively modern ideas of symbolic algebra could resolve questions from the age-old subject of Euclidean geometry.

The other thrust, somewhat more typical of Fermat, eventually turned out to be the more important. It began with an algebraic expression and used it to generate a geometrical figure on the plane, as we did with $y = x^2 + 1$ above and as Newton did when he graphed his six dozen cubics. Fermat had this approach in mind when he wrote, "Whenever in a final equation two unknown quantities are found, we have a locus, the extremity of one of these describing a line, straight or curved."[7] Fermat's insight, which Carl Boyer called "one of the most significant statements in the history of mathematics," allowed mathematicians to generate new curves at will simply by plotting points of ever more intricate equations.[8]

Prior to the appearance of analytic geometry, the supply of curves had been limited to those occurring "naturally." Mathematicians were acquainted with circles,

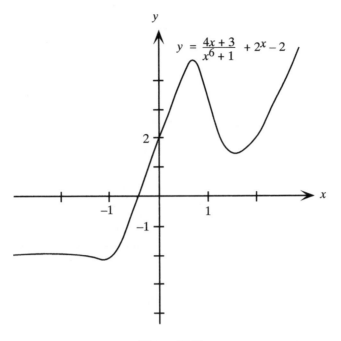

Figure XY.2

ellipses, and spirals, for these had their origins in well-known geometrical problems. But a curve like the graph of

$$y = \frac{4x + 3}{x^6 + 1} + 2^x - 2$$

in Figure XY.2 would have been utterly unimaginable. By cooking up strange equations, mathematicians generated curves twisting in heretofore unseen shapes across the x-y plane. And by acquiring this larger repertoire of curves, they gained insights that would prove essential to the development of calculus.

In the remainder of the chapter, we consider the two opposite but fundamental aspects of analytic geometry. We begin by examining how the geometry of a curve can guide us in understanding its algebraic properties.

Actually, we have seen examples of this phenomenon elsewhere in the book. Geometric diagrams motivated our discussion of differential and integral calculus and figured prominently in our development of Newton's method. An even simpler example came in Chapter D with Cardano's claim that no two real numbers could sum to 10 and multiply to 40. We verified his assertion using the maximizing techniques of differential calculus. But the whole matter could have been disposed of at once by looking at the graph of the product function $y = -x^2 + 10x$, which appeared as Figure D.9 and which we here reproduce as Figure XY.3.

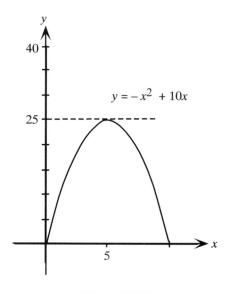

Figure XY.3

Once we have established that the products in question must be y-coordinates of points on this curve, it is obvious that their product cannot reach 40. The diagram makes it immediately clear that the largest product possible—that is, the highest point on the graph—is 25. Here restrictions on the product of the two numbers, a purely algebraic concern, are evident from the associated geometry of the curve.

Or we can look back to Chapter K where we observed that no algebraic technique provides an exact solution for an equation such as $x^7 - 3x^5 + 2x^2 - 11 = 0$. In such cases we touted Newton's method as a means to an approximate solution. But there is a different, albeit less efficient, route to the approximation, one that needs only input from analytic geometry.

We begin by graphing $y = x^7 - 3x^5 + 2x^2 - 11$. This, of course, is a horrendous problem to do by hand; the tedium of plotting points for such an equation would be forbidding indeed. However, technology makes it a snap. There is computer software, or even calculators that fit in the palm of your hand, that can graph something like this in seconds by plotting far more points than a human being would care to do in a month. This is what is illustrated in Figure XY.4.

Because we wish to solve $x^7 - 3x^5 + 2x^2 - 11 = 0$, we must find an x value that makes $y = 0$ in our equation above. This is simply the first coordinate of the so-called x-intercept where the graph cuts across the x-axis. A glance at Figure XY.4 indicates that there is only one such intercept, so our seventh-degree equation has only one solution. What we need to do is to zero in on this number.

Calculators and readily available software with a "zoom" feature allow us to do precisely that. It is as though we are putting a portion of the graph under a magnifying glass. We first identify a point toward which we shall zoom—in this case a little

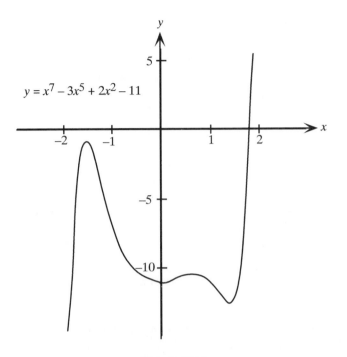

Figure XY.4

below $x = 2$ —and give the proper command. The result, shown in Figure XY.5, is an enlargement of the region around the intercept. From the geometry of the situation, it now appears that the solution lies somewhere near $x = 1.8$.

If that is too imprecise, so we zoom again. And we can keep this up until we are looking at a very tiny section of the x-axis—say, only 0.00001 unit across—within

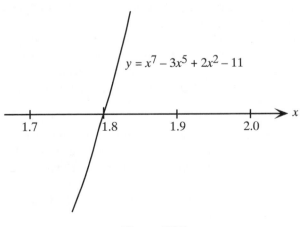

Figure XY.5

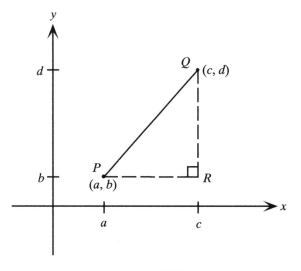

Figure XY.6

which the curve cuts the axis. In this manner we shall have approximated the x-intercept to a high degree of accuracy.

For the present example, we zoomed a few more times until zeroing in on $x = 1.7998295$ as an approximate solution. As a check, we substitute into the original equation to get

$$(1.7998295)^7 - 3(1.7998295)^5 + 2(1.7998295)^2 - 11 = -0.000004$$

so we are very close. Thanks to a graphing calculator, it was all fairly painless.

Before leaving this problem, we make two important observations. First, such a solution requires technology unthinkable a few decades ago but commonplace today. The hardware necessary to calculate and plot hundreds of points in a few seconds is the engineer's gift to mathematicians.

More importantly, although this was fundamentally an algebra problem—finding the solution of a seventh-degree equation—our solution was driven by the geometry of a curve crossing the x-axis. In such cases, mathematicians speak of the "power of visualization," and it is hard to argue that analytic geometry, augmented by the computer, holds great power indeed.

This has been an example of geometry in the service of algebra. Now we turn the tables and apply the weapons of algebra to prove a theorem of geometry. We shall need two preliminaries: the algebraic treatment of distance and of slope. Because we discussed the latter in Chapter D, we begin with a word about the former.

Suppose we are given points P and Q in Figure XY.6 and are asked to find the distance between them, which is the length of the solid line PQ. Letting the points have respective coordinates (a, b) and (c, d) as shown, we draw the dotted lines, thereby forming right triangle PQR. By the simple expedient of reading lengths off

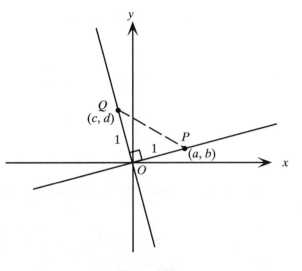

Figure XY.7

the x-axis, we see that $\overline{PR} = c - a$; a similar use of the y-axis shows that $\overline{QR} = d - b$. Then, by the Pythagorean theorem we have

$$\text{distance between } P \text{ and } Q = \sqrt{\overline{PR}^2 + \overline{QR}^2} = \sqrt{(c-a)^2 + (d-b)^2}$$

This, not surprisingly, is called the ***distance formula*** in analytic geometry.

As to the algebraic formulation of slope, we recall from Chapter D that the line between P and Q in Figure XY.6 has slope

$$m = \frac{\text{rise}}{\text{run}} = \frac{d-b}{c-a}$$

We issue a word of warning: If a line is vertical, then any two of its points have the same first coordinate, and this puts a zero in the denominator of the slope formula. The result is therefore not a real number, and we say that the slope of a vertical line is undefined. To avoid this troubling complication, we insist in all that follows that no line be vertical.

The concept of slope provides us with algebraic characterizations of parallelism and perpendicularity, two geometric concepts that show up early in Euclid's geometry. The intuitive idea of slope as "inclination" makes it readily apparent that lines are parallel if and only if they have the same slope. But an analogous characterization of perpendicularity is by no means obvious and deserves a brief discussion.

Suppose two lines meet at right angles. To move into the realm of analytic geometry, we superimpose the coordinate axes, putting the origin precisely at the point of intersection of the lines as depicted in Figure XY.7.

Along each line, take segments of length 1 terminating at points P with coordinates (a, b) and Q with coordinates (c, d). By the distance formula above,

$$\sqrt{(a-0)^2 + (b-0)^2} = \overline{OP} = 1, \text{ and so}$$

$$a^2 + b^2 = \left(\sqrt{a^2 + b^2}\right)^2 = 1^2 = 1$$

By the same reasoning,

$$c^2 + d^2 = \left(\sqrt{c^2 + d^2}\right)^2 = 1$$

and we conclude that

$$a^2 + b^2 + c^2 + d^2 = 1 + 1 = 2 \qquad\qquad (*)$$

Now draw the dotted line PQ to form right triangle POQ. The Pythagorean theorem guarantees that

$$\overline{PQ} = \sqrt{1^2 + 1^2} = \sqrt{2}$$

On the other hand, from the distance formula we get

$$\overline{PQ} = \sqrt{(c-a)^2 + (d-b)^2}$$

Equating these and squaring yields

$$(\sqrt{2})^2 = \left(\sqrt{(c-a)^2 + (d-b)^2}\right)^2, \text{ and so}$$
$$2 = (c-a)^2 + (d-b)^2 = c^2 - 2ac + a^2 + d^2 - 2bd + b^2$$
$$= (a^2 + b^2 + c^2 + d^2) - 2ac - 2bd$$
$$= 2 - 2ac - 2bd$$

by (*) above. Thus $2 = 2 - 2ac - 2bd \;\rightarrow\; 0 = -2ac - 2bd \;\rightarrow\; ac = -bd$.

The significance of this last equation becomes evident when we introduce slopes into the discussion. The slope of the line through O and P is

$$m_1 = \frac{b-0}{a-0} = \frac{b}{a}$$

and that of the line connecting O and Q is $m_2 = d/c$. Thus we have

$$m_1 \times m_2 = \frac{b}{a} \times \frac{d}{c} = \frac{bd}{ac}$$

$$= \frac{bd}{-bd} \text{ because we have just established that } ac = -bd$$

$$= -1.$$

Thus when two lines are perpendicular, the product of their slopes is -1. This may seem peculiar, but it is merely the Pythagorean theorem transposed into the world of analytic geometry.

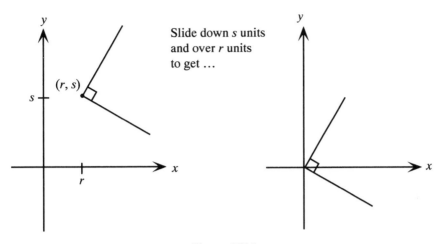

Figure XY.8

What if the lines intersect somewhere other than at the origin? For instance, consider the perpendicular lines meeting at the point (r, s) on the left side of Figure XY.8. We can rigidly slide the entire configuration down s units and over r units, thereby bringing the intersection point to $(0, 0)$ while maintaining the inclinations of the lines, as illustrated on the right side of the same figure. The result is a situation like that of Figure XY.7, so the previous discussion shows that the product of the slopes of the transferred lines is -1. But the original lines have the same inclinations as the transferred ones, so their slopes also multiply to -1.

With this as prologue, we are ready to prove a geometric theorem with algebraic tools. Our argument requires the distance formula, the notion of slope, and the characterizations of parallelism and perpendicularity—in short, all of the weapons we have assembled over the last few pages. The proposition is about a ***rhombus,*** which, it may be recalled, is a parallelogram all of whose sides are congruent.

THEOREM: If the diagonals of a parallelogram are perpendicular, the parallelogram is a rhombus.

PROOF: Begin with parallelogram $OABC$ shown in Figure XY.9. It has been oriented with vertex O at the origin and side OA running along the x-axis to the point $(a, 0)$ so that $\overline{OA} = a$. (Were we given a parallelogram not so favorably aligned, we could slide and rotate it—without changing the lengths or relative positions of the sides—so as to place it like that shown in Figure XY.9.)

Point C has coordinates (b, c), and because the figure is a parallelogram, side CB must run parallel to the x-axis. This guarantees that the second coordinates of C and B are the same, so we label B as (d, c). But sides OC and AB are also parallel. This means

$$\frac{c}{b} = \text{slope of line } OC = \text{slope of } AB = \frac{c}{d - a}$$

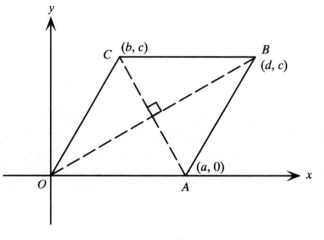

Figure XY.9

Cross-multiply to get $c(d-a)=bc$. From this it follows that $d-a=b$, or equivalently, $d=b+a$.

Now we have made an important assumption about this parallelogram: Its diagonals OB and AC are perpendicular. As we have seen, this means that the product of their slopes is -1. Thus we have

$$-1 = (\text{slope of } OB) \times (\text{slope of } AC) = \frac{c}{d} \times \frac{c}{b-a}$$

$$= \frac{c}{b+a} \times \frac{c}{b-a} \text{, because we showed that } d=b+a$$

$$= \frac{c^2}{b^2-a^2}$$

But because

$$\frac{c^2}{b^2-a^2} = -1$$

we see that $b^2 - a^2 = -c^2$, or simply that $a^2 = b^2 + c^2$. Then, by the distance formula,

$$\overline{OC} = \sqrt{(b-0)^2 + (c-0)^2} = \sqrt{b^2 + c^2}$$

$$= \sqrt{a^2} \text{ by the conclusion above}$$

$$= a = \overline{OA}$$

In short, OC and OA have the same length.

It is now easy to finish up. The length of CB is

$$\sqrt{(d-b)^2 + (c-c)^2} = \sqrt{(d-b)^2} = d-b = (b+a)-b = a$$

so *CB* and *OA* have the same length. Finally,

$$\overline{AB} = \sqrt{(d-a)^2 + (c-0)^2} = \sqrt{b^2 + c^2} = \overline{OC} = \overline{OA} = a$$

Consequently, all four sides of parallelogram *OABC* are equally long. It is a rhombus, and the proof is complete. ■

As the reader may sense, this result is quite easy to prove strictly within the confines of Euclidean geometry, where we employ the expected tools of congruent triangles, alternate interior angles of parallel lines, and so on. What is significant about the proof above is its algebraic nature. Taking Descartes's advice, we did "not hesitate to introduce these arithmetical terms into geometry" and thereby showed we had a rhombus by manipulating a string of equations.

There are examples of far greater significance that could have been given to bolster support for analytic geometry. For instance, this is a splendid arena for studying the so-called conic sections—the ellipse, parabola, and hyperbola. Within the framework of the *x-y* axes, these figures are much more easily understood than by treating them, in the Greek manner, as various cross-sections of cones.

However, the chapter has run its course. Besides, if the conic sections were discussed here, we "should deprive you of the pleasure of mastering it yourself." Instead, we end with a last salute to the analytic geometry of the *x-y* plane, the fusion of geometry and algebra that remains the happiest marriage in all of mathematics.

$e^{i\pi} + 1 = 0$

The title of this final chapter suggests that we are at a loss for words. Actually the choice of a single-letter title is perfectly appropriate, for this is a section about complex numbers and these, in the mathematician's alphabet, are invariably denoted by the letter z.

Our goal is to describe what complex numbers are, where they came from, and why they play such a dominant role in modern mathematics. Theirs has been a tortured history, one requiring the abandonment of widely held prejudices about the meaning of "number." According to Greek mythology, Athena burst full-grown from the head of her father, Zeus, but, as we shall see, imaginary numbers emerged by fits and starts from the heads of many mathematicians over many centuries.

To understand what was so troubling about this concept, we must look to the properties of familiar real numbers. As everyone knows, nonzero numbers come in two varieties, positive and negative, with zero fitting into neither category. Their arithmetic is governed by various rules, but for our discussion one stands out: The product of two positive numbers or the product of two negative numbers must be positive. For instance, $3 \times 4 = 12$, and $(-3) \times (-4) = 12$ as well.

Suppose, then, that we are looking for the square root of a negative number, such as $\sqrt{-15}$. This presents (dare we say it?) *real* trouble. Any number, when squared, is positive or zero; so no number, when squared, equals -15. As our mathematical forerunners observed, quantities such as $\sqrt{-15}$ should be readily dismissed as figments of the imagination.

Yet it is hard to keep a good idea down. This was apparent in the work of the sixteenth century Italian algebraists who made significant contributions to the art of equation solving. Along the way they inadvertently stumbled upon the square roots of negatives.

As we mentioned in Chapter Q, mathematicians of this period were still troubled by the concept of *negative* numbers, let alone their square roots. We see this aversion in Gerolamo Cardano's treatment of quadratic equations in his *Ars Magna* from 1545. For instance, he could handle the case called "square plus cosa equals number" (*cosa* is the unknown to be determined). In modern notation, this amounts to an equation such as $x^2 + 3x = 40$. But he refrained from considering "square minus cosa equals number." That is, he never examined equations such as $x^2 - 3x = 40$. Such an equation, involving a negative quantity, was tainted.

Instead Cardano described a procedure for solving $x^2 = 3x + 40$. From our perspective this is *equivalent* to $x^2 - 3x = 40$ and requires no separate case. But in a century when negative quantities were viewed so negatively, such algebraic gyrations were deemed unavoidable.

Those who do not believe in unicorns find it absurd to talk about their eating habits. Likewise those who doubted the existence of negative numbers surely regarded their *square roots* as doubly preposterous. Yet it was Cardano who took the first tentative step in this direction with a problem he posed in the *Ars Magna:* "someone says to you, divide 10 into two parts, one of which multiplied into the other shall produce 40."[1]

Cardano noted that the solution is impossible. Indeed, there are no *real* numbers with this property, as we proved using differential calculus in Chapter D. "Nevertheless," he observed, "we shall solve it in this fashion" and gave as solutions

$$5 + \sqrt{-15} \text{ and } 5 - \sqrt{-15}$$

Are these answers reasonable? First of all, their sum is

$$5 + \sqrt{-15} + 5 - \sqrt{-15} = 5 + 5 = 10$$

so that he did indeed "divide 10 into two parts." To determine the product of these parts, we apply the familiar multiplication rule

$$(a + b)(c + d) = ac + ad + bc + bd$$

where each term in the first parenthesis multiplies each term in the second. This gives

$$(5 + \sqrt{-15}) \times (5 - \sqrt{-15}) = 25 - 5\sqrt{-15} + 5\sqrt{-15} - (\sqrt{-15})^2$$
$$= 25 - (-15) = 25 + 15 = 40$$

because $(\sqrt{-15})^2 = -15$. So, Cardano seems to have found a way of dividing 10 into two parts whose product is 40.

But were his solutions "numbers"? What exactly *is* $\sqrt{-15}$? Cardano was more confused than convinced by his own answer, which he called "puzzling" and which he characterized as being "as subtle as it is useless," hardly a rousing endorsement for a new mathematical concept.[2]

Square roots of negatives could probably have been dismissed altogether were it not for the greatest achievement of the Italian mathematicians, the algebraic solution of the cubic equation. In *Ars Magna*, Cardano first published the solution to "cube plus cosa equals number"—that is, to a cubic of the form $x^3 + mx = n$, where m and n are real. He attacked the problem geometrically by decomposing three-dimensional cubes into various pieces, but in modern algebraic notation his solution is

$$x = \sqrt[3]{\frac{n}{2} + \sqrt{\frac{n^2}{4} + \frac{m^3}{27}}} \; - \; \sqrt[3]{-\frac{n}{2} + \sqrt{\frac{n^2}{4} + \frac{m^3}{27}}}$$

This formula works nicely in solving, for instance, the cubic $x^3 + 24x = 56$. Here $m = 24$ and $n = 56$, and we compute

$$\sqrt{\frac{n^2}{4} + \frac{m^3}{27}} = \sqrt{\frac{56^2}{4} + \frac{24^3}{27}} = \sqrt{784 + 512} = \sqrt{1{,}296} = 36$$

Then, according to Cardano's formula above, the cubic's solution is

$$x = \sqrt[3]{\frac{56}{2} + 36} \; - \; \sqrt[3]{-\frac{56}{2} + 36}$$

$$= \sqrt[3]{28 + 36} \; - \; \sqrt[3]{-28 + 36} = \sqrt[3]{64} \; - \; \sqrt[3]{8} = 4 - 2 = 2$$

Indeed, $x = 2$ satisfies the equation, because $2^3 + 24(2) = 8 + 48 = 56$ as required. Everything is working fine.

But what does one make of the cubic $x^3 - 78x = 220$? Here we have $m = -78$ and $n = 220$, so that

$$\sqrt{\frac{n^2}{4} + \frac{m^3}{27}} = \sqrt{\frac{220^2}{4} + \frac{(-78)^3}{27}} = \sqrt{12{,}100 - 17{,}576} = \sqrt{-5{,}476}$$

We have encountered the dreaded square root of a negative. Cardano's wonderful cubic formula has led to trouble.

What really gave mathematicians headaches was that the cubic $x^3 - 78x = 220$ *does* have a real solution, namely $x = 10$. (It is easy to check that $10^3 - 78(10) = 1{,}000 - 780 = 220$.) Moreover, we can find two other real solutions: $-5 + \sqrt{3}$ and $-5 - \sqrt{3}$. The situation is most unsatisfactory, for here is a real cubic with three real solutions, yet Cardano's formula seemed incapable of finding any of them. Scholars were more puzzled than ever.

So matters stood for a generation until another Italian mathematician, Rafael Bombelli (ca. 1526–1572), had a striking insight in his *Algebra* of 1572. He suggested that square roots of negatives could be introduced, at least temporarily, in going from the real cubic to its real solutions. In this fashion these strange and troubling numbers would serve as an intermediate tool for solving cubics.

To see what Bombelli had in mind, we return to the point at which trouble appeared in the previous example. Temporarily suspending any prejudice against square roots of negatives, we write

$$\sqrt{-5{,}476} = \sqrt{5{,}476 \times (-1)} = \sqrt{5{,}476} \times \sqrt{-1} = 74\sqrt{-1}$$

because $\sqrt{5{,}476} = 74$. Then applying Cardano's formula to the cubic $x^3 - 78x = 220$ gives

$$
\begin{aligned}
x &= \sqrt[3]{\frac{n}{2} + \sqrt{\frac{n^2}{4} + \frac{m^3}{27}}} \; - \; \sqrt[3]{-\frac{n}{2} + \sqrt{\frac{n^2}{4} + \frac{m^3}{27}}} \\[2mm]
&= \sqrt[3]{\frac{220}{2} + \sqrt{-5{,}476}} \; - \; \sqrt[3]{-\frac{220}{2} + \sqrt{-5{,}476}} \\[2mm]
&= \sqrt[3]{110 + 74\sqrt{-1}} \; - \; \sqrt[3]{-110 + 74\sqrt{-1}} \qquad\qquad (*)
\end{aligned}
$$

This seems to have made matters worse, for we not only retain the square root of -1 but now find it embedded within a cube root. However, Bombelli recognized that this expression, if approached gingerly, could yet do the job.

We need a few computations to see why. First, note that

$$
\begin{aligned}
(5 + \sqrt{-1})\,(5 + \sqrt{-1}) &= 5^2 + 5\sqrt{-1} + 5\sqrt{-1} + \sqrt{-1}\,^2 \\
&= 25 + 10\sqrt{-1} + (-1) = 24 + 10\sqrt{-1}
\end{aligned}
$$

where we have used the fact that $\sqrt{-1}\,^2 = -1$. Put another way, we have shown that $(5 + \sqrt{-1})^2 = 24 + 10\sqrt{-1}$. We push the expansion a step further and compute:

$$
\begin{aligned}
(5 + \sqrt{-1})^3 &= (5 + \sqrt{-1}) \times (5 + \sqrt{-1})^2 = (5 + \sqrt{-1}) \times (24 + 10\sqrt{-1}) \\
&= 120 + 50\sqrt{-1} + 24\sqrt{-1} + 10(\sqrt{-1})^2 = 120 + 74\sqrt{-1} + 10(-1) \\
&= 120 + 74\sqrt{-1} - 10 = 110 + 74\sqrt{-1}
\end{aligned}
$$

This should stir the mathematician's heart, for it is precisely the expression appearing under the first of the two cube roots in the solution labeled (*) above. Having thus found that

$$(5 + \sqrt{-1})^3 = 110 + 74\sqrt{-1}$$

we take cube roots of each side to conclude

$$5 + \sqrt{-1} = \sqrt[3]{110 + 74\sqrt{-1}}$$

A similar computation shows that $(-5 + \sqrt{-1})^3 = -110 + 74\sqrt{-1}$, and so

$$-5 + \sqrt{-1} = \sqrt[3]{-110 + 74\sqrt{-1}}$$

At last we are able to make sense of Cardano's formula. Upon returning to (*) and substituting the cube roots just determined, we find

$$x = \sqrt[3]{110 + 74\sqrt{-1}} - \sqrt[3]{-110 + 74\sqrt{-1}}$$
$$= (5 + \sqrt{-1}) - (-5 + \sqrt{-1})$$
$$= 5 + \sqrt{-1} + 5 - \sqrt{-1} = 10$$

As we previously checked, $x = 10$ *is* a solution to the original cubic. Strangely, $\sqrt{-1}$ has come to the rescue.

At this point a serious objection can be raised: How would anyone know at the outset that $5 + \sqrt{-1}$ was the cube root of $110 + 74\sqrt{-1}$? It is certainly not self-evident, and Bombelli himself had to rely on contrived examples (such as this one) where he knew the identity of the cube root beforehand. As to how to find the cube root of the general expression $a + b\sqrt{-1}$, he had no clue, and this would remain a mystery for some time.

Bombelli's approach—"a wild thought," he called it—appeared to work as much by magic as by logic. "The whole matter," he wrote, "seems to rest on sophistry rather than on truth."[3] Still, his willingness to introduce square roots of negatives was a significant step. It allowed the cubic to be solved; it salvaged Cardano's formula; and it put a new kind of number under the mathematical spotlight.

As might be expected for so troubling an idea, acceptance was neither universal nor immediate. Six decades after Bombelli, Descartes coined a term for numbers such as $\sqrt{-9}$ in this passage from his *Géométrie:* "Neither the true nor the false [negative] roots are always real; sometimes they are imaginary."[4] To label a mathematical concept as imaginary—as a gnome is imaginary or the Mad Hatter is imaginary—is to suggest that it is somehow hypothetical, paradoxical, or hallucinatory. In spite of such a connotation, his terminology has persisted down to the present day.

Later in the seventeenth century, Newton rendered a verdict of sorts when he referred to these numbers as "impossible."[5] Meanwhile Leibniz adopted a pseudobiological viewpoint in describing "that amphibian between being and non-being, which we call the imaginary root of negative unity."[6] Comparing imaginary numbers to amphibians may be better than comparing them to the Mad Hatter, but not much.

These numbers remained second-class citizens well into the eighteenth century. By then certain questions arising from calculus, along with the penetrating insight of Leonhard Euler, moved imaginary numbers toward full partnership in the mathematical enterprise. And it was he who helped standardize the symbol i for $\sqrt{-1}$.

Using this notation, we today define a ***complex number*** to be one of the form $z = a + bi$, where a and b are real. (Note at last the appearance of letter from the chapter's title.) Examples of complex numbers are $3 + 4i$ or $2 - 7i$. Because either a or b may be zero, the purely imaginary number $i(= 0 + 1i)$ and any real number $a(= a + 0i)$ fall under the broader heading of complex number. In this sense, the complex numbers contain within them all the number systems we met in Chapter Q.

Euler did much more than provide a bit of notation. He plugged a logical gap that had troubled his predecessors by showing how to find the cube root—or indeed any root—of a general quantity $a + bi$ and in the process demonstrated that a nonzero complex number has two different square roots, three different cube roots, four different fourth roots, and so on. For instance, the real number 8 obviously has as one of its cube roots the real number 2. Less obvious is that $-1 + i\sqrt{3}$ and $-1 - i\sqrt{3}$ are two additional cube roots of 8. (The skeptical reader is advised to cube these complex numbers and observe that the result is 8.)

Euler also investigated powers of complex numbers. It is easy to see that $i^2 = \sqrt{-1}\,^2 = -1$ and $i^3 = i^2 \times i = (-1) \times i = -i$. But Euler was after bigger game. With his customary boldness, he proved the remarkable fact that

$$e^{i\pi} + 1 = 0$$

As any mathematician is quick to observe, no other equation is quite like this one, for it provides a link among the most important constants in all of mathematics. Not only are 0 and 1 featured players, but π (from Chapter C), e (from Chapter N), and i (from Chapter Z) return for a final curtain call. This is truly an all-star cast.

Even stranger was Euler's evaluation of i^i, an imaginary power of an imaginary number. It seems preposterous even to think about such a thing, but all we need are Euler's formula above and two familiar rules of exponents:

$$(a^r)^s = a^{rs} = (a^s)^r \text{ and } a^{-r} = 1/a^r$$

Assuming (purely on faith) that these rules apply even if the base and/or exponent are complex numbers, we reason as follows:

$e^{i\pi} + 1 = 0$ implies that $e^{i\pi} = -1$, which in turn means that $e^{i\pi} = i^2$

Raise each side of the last equation to the power i to get $(e^{i\pi})^i = (i^2)^i$ and then apply the first exponent rule to convert this to

$$e^{i2\pi} = (i^i)^2$$

But again $i^2 = -1$, so we have $e^{-\pi} = (i^i)^2$. Take square roots of each side to get

$$\sqrt{e^{-\pi}} = i^i$$

Finally, because $e^{-\pi}$ is the same as $1/e^\pi$, we conclude

$$i^i = \frac{1}{\sqrt{e^\pi}}$$

Note that in this case an imaginary power of an imaginary number turned out to be the *real number*

$$\frac{1}{\sqrt{e^\pi}}$$

This is weird. A century later, U.S. logician Benjamin Peirce summarized most people's reaction to Euler's discovery when he said of this strange result, "we have not the slightest idea of what this equation means, but we may be sure that it means something very important."[7]

Euler deserves much of the credit for popularizing complex numbers. He showed how to find their powers and roots and even defined such things as their logarithms. In a sense, he established their arithmetic and algebraic legitimacy.

But there was further to go. During the following century, the *calculus* of complex functions was developed by many mathematicians, chief among them Augustin-Louis Cauchy of France (1789–1857). With these innovations, mathematicians could find such things as the derivative of $z^3 + 4z - 2i$ or an integral such as

$$\int \frac{e^{iz}}{i}\, dz$$

Complex numbers had certainly come a long way.

But there is one final and overarching result that gives complex numbers their special status. It is called the fundamental theorem of algebra, and it establishes the algebraic superiority of complex numbers over the number systems that came before. As such, it is surely one of the great theorems of mathematics.

The proof of the fundamental theorem of algebra lies far beyond the limits of this book, but we can describe what it says and why it is important. Not surprisingly for a theorem with the word *algebra* in its name, it concerns the solution of equations.

Think back to Chapter A, where we discussed the natural numbers: 1, 2, 3, ... By anyone's standards, these are the simplest—the least complex—numbers around, and simplicity is surely part of their fascination and charm. At the same time, it renders them inadequate for equation solving.

Suppose, for instance, we wished to solve $2x + 3 = 11$. Its coefficients are 2, 3, and 11, so this equation is written entirely within the system of whole numbers. Moreover, its solution is $x = 4$, again a whole number. For this example the natural numbers are sufficient both to generate the equation and to furnish its solution.

But what about $2x + 11 = 3$? Although written within the system of whole numbers, this equation has no whole number solution. For even when we assign to x the smallest possible candidate, $x = 1$, the expression $2x + 11$ becomes $2(1) + 11 = 13$, which far exceeds the value of 3 that is sought. We here have an equation expressed in terms of whole numbers but having no whole number solution. In this sense, the natural numbers are algebraically incomplete.

Nor is the system of real numbers without defect. Consider the quadratic $x^2 + 15 = 0$, which has only real numbers (in fact, has only integers) as coefficients. Yet its solution, $x = \sqrt{-15}$, certainly is not real. Here, an equation written within the system of real numbers has no solution within the system. The reals are incomplete as well.

But the complex numbers exhibit no such deficiency. Such an algebraic escape is impossible. *This* is the essence of the fundamental theorem of algebra. It guarantees that for a polynomial equation whose coefficients are complex numbers, the solu-

Carl Friedrich Gauss on a German 10-mark note

tions must exist as complex numbers. This is true for a first-degree equation such as $3x + 8 = 2 + 3i$, which has the single complex solution $x = -2 + i$, or for a second degree equation such as $x^2 + x = 11 + 7i$, which has the two complex solutions $x = 3 + i$ and $x = -4 - i$. But the theorem also applies to a fifth-degree polynomial equation such as

$$5x^5 + ix^4 - 3x^3 + (8 - 2i)x^2 - 17x - i = 0$$

which must have five (perhaps repeated) solutions, all of them complex numbers. In fact, the degree of the polynomial does not matter. The ***fundamental theorem of algebra*** states that any nth degree polynomial equation written in the complex numbers will have n (perhaps repeated) solutions, all of which must themselves be complex numbers.

We should observe that the theorem does not provide an explicit way to *find* these complex solutions but merely proves that they exist. Still, it is a very significant and powerful result, for it shows that the system of complex numbers is adequate to provide solutions to *any* polynomial equation written within it.

A number of eighteenth-century mathematicians, Euler included, believed the theorem to be true but were unable to furnish a satisfactory proof.[8] This had to await the coming of Carl Friedrich Gauss, a mathematician who has appeared time and again in these pages. In Chapter A, Gauss was introduced as one of history's foremost number theorists, so it is perhaps fitting that we return to him at book's end. His 1799 doctoral dissertation from the University of Helmstadt contained the first proof of the fundamental theorem of algebra. This dissertation, resolving so important a question, is regarded as the greatest mathematical doctorate of all time. Its existence should keep other Ph.D. recipients suitably humble.

Cardano regarded the imaginary numbers as "useless," Leibniz placed these "amphibia" somewhere between reality and fiction, and Euler pried loose some of their most bizarre and intriguing properties. But it was Gauss who established that the complex numbers are the ideal system in which to solve equations. In a very real sense, the fundamental theorem establishes complex numbers as the algebraist's paradise.

AFTERWORD

The alphabet has been exhausted. At this point many readers may feel the same. Our mathabetical excursion began in Chapter A with the fundamental theorem of arithmetic. Midway through the book, in Chapter L, we encountered the fundamental theorem of calculus. And we have ended with the fundamental theorem of algebra.

Mixed among these fundamentals have been mathematics and mathematicians, diagrams and formulas, issues and controversies. While going from A to Z, we traveled from China to Cambridge, from Thales to the modern computer. Surely each chapter could have been expanded a hundredfold, but the limitations of space kept us from tarrying too long on any one subject. Some chapters, perhaps, could have been jettisoned altogether, but the interests of the author dictated that they stay.

In the final analysis this has been one individual's journey and nothing more. I thank the reader for coming along.

NOTES

Preface

1. Ann Hibler Koblitz, *A Convergence of Lives*, Birkhäuser, Boston, 1983, p. 231.

2. Proclus, *A Commentary on the First Book of Euclid's Elements*, trans. Glenn R. Morrow, Princeton U. Press, Princeton, NJ, 1970, p. 17.

Arithmetic

1. Morris Kline, *Mathematical Thought from Ancient to Modern Times*, Oxford U. Press, New York, 1972, p. 979.

2. David Wells, *The Penguin Dictionary of Curious and Interesting Numbers*, Penguin, New York, p. 257.

3. Florian Cajori, *A History of Mathematics*, Chelsea (reprint), New York, 1980, p. 167.

4. David Burton, *Elementary Number Theory*, Allyn and Bacon, Boston, 1976, p. 226.

5. *Focus*, newsletter of the Mathematical Association of America, Vol. 12, No. 3, June 1992, p. 3.

6. Leonard Eugene Dickson, *History of the Theory of Numbers*, Vol. 1, G. E. Stechert and Co., New York, 1934, p. 424.

7. Thomas L. Heath, *The Thirteen Books of Euclid's Elements*, Vol. 1, Dover, New York, 1956, pp. 349–350.

8. Donald J. Albers, Gerald L. Alexanderson, and Constance Reid, *More Mathematical People*, Harcourt Brace Jovanovich, Boston, 1990, p. 269.

9. Paul Hoffman, "The Man Who Loves Only Numbers," *The Atlantic Monthly*, November 1987, p. 64.

10. Ibid., p. 65.

11. Caspar Goffman, "And What Is Your Erdös Number?" *The American Mathematical Monthly*, Vol. 76, No. 7, 1969, p. 791.

Bernoulli Trials

1. David Eugene Smith, *A Source Book in Mathematics*, Dover, New York, 1959, p. 90.

2. Kline, *Mathematical Thought*, p. 473.

3. Charles C. Gillispie, ed., *Dictionary of Scientific Biography*, Vol. 2, Scribner's, New York, 1970, Johann Bernoulli, p. 53.

4. Anders Hald, *A History of Probability and Statistics and Their Applications before 1750*, Wiley, New York, 1990, p. 223.

5. Gillispie, *Dictionary of Scientific Biography*, Jakob Bernoulli, p. 50.

6. James R. Newman, *The World of Mathematics*, Vol. 3, Simon and Schuster, New York, 1956, pp. 1452–1453.

7. Hald, *History of Probability*, p. 257.

8. Gerd Gigerenzer et al., *The Empire of Chance*, Cambridge U. Press, New York, 1990, p. 29.

9. Newman, *World of Mathematics*, p. 1455.

10. Ibid., p. 1454.

11. Ian Hacking, *The Emergence of Probability*, Cambridge U. Press, New York, 1975, p. 168.

Circle

1. Vitruvius, *On Architecture*, trans. Frank Granger, Vol. 2, Loeb Classical Library, Cambridge, MA, 1962, p. 205.

2. Howard Eves, *An Introduction to the History of Mathematics*, 5th ed., Saunders, New York, 1983, p. 89.

3. Richard Preston, "Mountains of Pi," *The New Yorker*, March 2, 1992, pp. 36–67.

4. David Singmaster, "The Legal Values of Pi," *The Mathematical Intelligencer*, Vol. 7, No. 2, 1985, pp. 69–72.

5. Ibid., p. 69.

6. Ibid., p. 70.

Differential Calculus

1. Dirk Struik, ed., *A Source Book in Mathematics: 1200–1800*, Princeton U. Press, Princeton, NJ, 1986, pp. 272–273.

2. James Stewart, *Calculus*, 2nd ed., Brooks/Cole, Pacific Grove, CA, 1991, p. 56.

Euler

1. C. Boyer and Uta Merzbach, *A History of Mathematics*, 2nd ed., Wiley, New York, 1991, p. 440.

2. G. Waldo Dunnington, *Carl Friedrich Gauss: Titan of Science*, Exposition Press, New York, 1955, p. 24.

3. Carl Boyer, *History of Analytic Geometry*, Scripta Mathematica, New York, 1956, p. 180.

4. Dunnington, *Carl Friedrich Gauss: Titan of Science*, pp. 27–28.

5. G. G. Joseph, *The Crest of the Peacock*, Penguin, New York, 1991, p. 323.

6. "Glossary," *Mathematics Magazine*, Vol. 56, No. 5, 1983, p. 317.

7. E. H. Taylor and G. C. Bartoo, *An Introduction to College Geometry*, Macmillan, New York, 1949, pp. 52–53.

8. André Weil, *Number Theory: An Approach through History*, Birkhäuser, Boston, 1984, p. 261.

9. W. Dunham, *Journey through Genius*, Wiley, New York, 1990, Chapter 9.

10. Weil, *Number Theory*, p. 277.

Fermat

1. Weil, *Number Theory*, p. 39.

2. E. T. Bell, *The Last Problem*, (Introduction and Notes by Underwood Dudley), Mathematical Association of America, Washington, DC, 1990, p. 265.

3. Boyer and Merzbach, *History of Mathematics*, p. 344.

4. Ibid, p. 333.

5. Weil, *Number Theory*, p. 51.

6. Michael Sean Mahoney, *The Mathematical Career of Pierre de Fermat*, Princeton U. Press, Princeton, NJ, 1973, p. 311.

7. Burton, *Elementary Number Theory*, p. 264.

8. Smith, *Source Book in Mathematics*, p. 213.

9. Ibid.

10. Harold M. Edwards, *Fermat's Last Theorem*, Springer-Verlag, New York, 1977, p. 73.

11. Bell, *Last Problem*, p. 300.

12. Gina Kolata, "At Last, Shout of 'Eureka!' in Age-Old Math Mystery," *New York Times*, June 24, 1993, p. 1; Michael Lemonick, "*Fini* to Fermat's Last Theorem," *Time*, July 5, 1993, p. 47.

13. Edwards, *Fermat's Last Theorem*, p. 38.

Greek Geometry

1. Ivor Thomas, *Greek Mathematical Works*, Vol. 1, Loeb Classical Library, Cambridge, MA, 1967, pp. viii–ix.

2. Ibid., p. 391.

3. Ibid., p. 147.

4. Heath, *Thirteen Books of Euclid's Elements*, Vol. 1, p. 153.

5. Dunham, *Journey through Genius*, pp. 37–38.

6. Thomas, *Greek Mathematical Works*, p. ix.

7. Heath, *Thirteen Books of Euclid's Elements*, Vol. 1, pp. 253–254.

8. Proclus, *Commentary on the First Book*, p. 251.

9. Ibid.

10. A. Conan Doyle, *The Complete Sherlock Holmes*, Garden City Books, Garden City, NY, 1930, p. 12.

11. Heath, *Thirteen Books of Euclid's Elements*, Vol. 1, p. 4.

12. *American Mathematical Monthly*, Vol. 99, No. 8, October 1992, p. 773.

13. Morris Kline, *Mathematics in Western Culture*, Oxford U. Press, New York, 1953, p. 54.

14. G. H. Hardy, *A Mathematician's Apology*, Cambridge U. Press, New York, 1967, pp. 80–81.

Hypotenuse

1. Elisha Scott Loomis, *The Pythagorean Proposition*, National Council of Teachers of Mathematics, Washington, DC, 1968.

2. Frank J. Swetz and T. I. Kao, *Was Pythagoras Chinese?* Pennsylvania State U. Press, University Park, PA, 1977, pp. 12–16.

3. Edmund Ingalls, "George Washington and Mathematics Education," *Mathematics Teacher*, Vol. 47, 1954, p. 409.

4. James Mellon, ed., *The Face of Lincoln*, Viking, New York, 1979, p. 67.

5. Ulysses S. Grant, *Personal Memoirs*, Bonanza Books, New York (facsimile of 1885 ed.), pp. 39–40.

6. *The New England Journal of Education*, Vol. 3, Boston, 1876, p. 161.

7. *The Inaugural Addresses of the American Presidents*, annotated by Davis Newton Lott, Holt, Rinehart & Winston, New York, 1961, p. 146.

8. *The New England Journal of Education*, Vol. 3, Boston, 1876, p. 161.

Isoperimetric Problem

1. Virgil, *The Aeneid*, trans. Rolfe Humphries, Scribner's, New York, 1951, p. 16.

2. Proclus, *Commentary on the First Book*, p. 318.

3. Thomas, *Greek Mathematical Works*, Vol. 2, p. 395.

4. Ibid., pp. 387–389.

5. Ibid., p. 589.

6. Ibid., p. 593.

Justification

1. Michael Atiyah comment in "A Mathematical Mystery Tour," *Nova*, PBS television program.

2. Bertrand Russell, *The Basic Writings of Bertrand Russell: 1903–1959*, Robert Egner and Lester Denonn, eds., Simon and Schuster, New York, 1961, p. 175.

3. Charles Darwin, *The Autobiography of Charles Darwin*, Dover Reprint, New York, 1958, p. 55.

4. Boyer, *History of Analytic Geometry*, p. 103.

5. Thomas, *Greek Mathematical Works*, Vol. 1, p. 423.

6. Russell, *The Basic Writings of Bertrand Russell: 1903–1959*, p. 163.

7. John Bartlett, ed., *Familiar Quotations*, Little, Brown, Boston, 1980, p. 746.

8. Barry Cipra, "Solutions to Euler Equation," *Science*, Vol. 239, 1988, p. 464.

9. Hardy, *Mathematician's Apology*, p. 94.

10. Malcolm Browne, "Is a Math Proof a Proof If No One Can Check It?" *New York Times*, December 20, 1988, p. 23.

Knighted Newton

1. John Fauvel, Raymond Flood, Michael Shortland, and Robin Wilson, *Let Newton Be!*, Oxford U. Press, New York, 1988, pp. 11–12.

2. Ibid., p. 14.

3. Kline, *Mathematics in Western Culture*, p. 214.

4. Adolph Meyer, *Voltaire: Man of Justice*, Howell, Soskin Publishers, New York, 1945, p. 184; Fauvel et al., *Let Newton Be!*, p. 185.

5. R. S. Westfall, *Never at Rest*, Cambridge U. Press, New York, 1980, p. 270.

6. Ibid., pp. 273–274.

7. Ibid., p. 266.

8. Ibid., p. 202.

9. Joseph E. Hoffman, *Leibniz in Paris*, Cambridge U. Press, Cambridge, UK, 1974, p. 229.

10. Westfall, *Never at Rest*, pp. 715–716.

11. Ibid., p. 761.

12. Derek Whiteside, ed., *The Mathematical Papers of Isaac Newton*, Vol. 2, Cambridge U. Press, Cambridge, UK, 1968, pp. 221–223.

Lost Leibniz

1. J. M. Child, *The Early Mathematical Manuscripts of Leibniz*, Open Court Publishing, London, 1920, p. 11.

2. J. Hoffman, *Leibniz in Paris*, pp. 2–3.

3. Ibid., p. 15.

4. C. H. Edwards, Jr., *The Historical Development of Calculus*, Springer-Verlag, New York, 1979, p. 234.

5. Child, *Early Mathematical Manuscripts of Leibniz*, p. 12.

6. J. Hoffman, *Leibniz in Paris*, pp. 91–93.

7. Ibid., p. 151.

8. Eves, *Introduction to History of Mathematics*, p. 309.

Mathematical Personality

1. G. Pólya, "Some Mathematicians I Have Known," *American Mathematical Monthly*, Vol. 76, No. 7, 1969, pp. 746–753.

2. Paul Halmos, *I Have a Photographic Memory*, American Mathematical Society, Providence, RI, 1987, p. 2.

3. Pólya, "Some Mathematicians I Have Known," pp. 746–753.

4. Westfall, *Never at Rest*, p. 192.

5. Eves, *Introduction to History of Mathematics*, p. 370.

6. John F. Bowers, "Why Are Mathematicians Eccentric?" *New Scientist* 22/29, December 1983, pp. 900–903.

7. Westfall, *Never at Rest*, p. 105.

8. Harold Taylor and Loretta Taylor, *George Pólya: Master of Discovery*, Dale Seymour Publications, Palo Alto, CA, 1993, p. 21.

9. Ed Regis, *Who Got Einstein's Office*, Addison-Wesley, Reading, MA, 1987, p. 195.

10. Scott Rice, ed., *Bride of Dark and Stormy*, Penguin, New York, 1988, p. 124.

11. *Math Matrix* (newsletter of the Department of Mathematics, The Ohio State University), Vol. 1, No. 5, 1986, p. 3.

12. Don Albers and G. L. Alexanderson, "A Conversation with Ivan Niven," *College Mathematics Journal*, Vol. 22, No. 5, November 1991, p. 394.

13. JoAnne Growney, "Misunderstanding," *Intersections*, Kadet Press, Bloomsburg, PA, 1993, p. 48.

Natural Logarithm

1. Leonhard Euler, *Opera Omnia*, Vol. 8, Ser. 1, B. G. Teubneri, Leipzig, 1922, p. 128.

2. Charles Darwin, *The Autobiography of Charles Darwin*, Dover, New York, 1958, pp. 42–43.

Origins

1. Victor Katz, *A History of Mathematics: An Introduction*, HarperCollins, New York, 1993, p. 4.

2. Joseph, *Crest of the Peacock*, p. 61.

3. Ibid., p. 80.

4. Ibid., p. 82.

5. Ibid., pp. 83–84.

6. Swetz and Kao, *Was Pythagoras Chinese?*, p. 29.

7. Boyer and Merzbach, *History of Mathematics*, p. 223.

8. Cajori, *History of Mathematics*, p. 87.

9. Harry Carter, trans., *The Histories of Herodotus*, Vol. 1, Heritage Press, New York, 1958, p. 131.

Prime Number Theorem

1. Boyer and Merzbach, *History of Mathematics*, p. 501.

Quotient

1. René Descartes, *The Geometry of René Descartes*, trans. David Eugene Smith and Marcia L. Latham, Dover, New York, 1954, p. 2.

2. Kline, *Mathematical Thought*, 1972, pp. 592–593.

3. Ibid., p. 251.

4. Ibid., pp. 593–594.

5. Ibid., p. 981.

Russell's Paradox

1. Ronald W. Clark, *The Life of Bertrand Russell*, Knopf, New York, 1976, p. 7.

2. Ibid., p. 28.

3. Robert E. Egner and Lester E. Denonn, eds., *The Basic Writings of Bertrand Russell 1903–1959*, Simon & Schuster, New York, 1961, p. 253.

4. Ibid., pp. 253–254.

5. Bertrand Russell, *Introduction to Mathematical Philosophy*, Macmillan, New York, 1919, p. 1.

6. Bertrand Russell, *The Autobiography of Bertrand Russell, 1872–1914,* George Allen & Unwin Ltd., London, 1967, p. 145.

7. "A Mathematical Mystery Tour," *Nova*, PBS television program.

8. Russell, *Autobiography*, p. 152.

9. Clark, *Life of Russell*, p. 258.

10. Egner and Denonn, *Basic Writings*, p. 595.

11. Ibid., p. 589.

12. Ibid., p. 253.

13. A. J. Ayer, *Bertrand Russell*, U. of Chicago Press, Chicago, 1972, p. 17.

14. Clark, *Life of Russell,* p. 53.

15. Ibid., p. 441.

16. Ibid., p. 334.

17. Egner and Denonn, *Basic Writings,* p. 479.

18. Clark, *Life of Russell,* p. 382.

19. Egner and Denonn, *Basic Writings*, p. 352.

20. Ibid., p. 298.

21. Clark, *Life of Russell*, p. 451.

22. Egner and Denonn, *Basic Writings*, p. 63.

23. Clark, *Life of Russell*, p. 202.

24. Bertrand Russell, *My Philosophical Development*, George Allen & Unwin Ltd., London, 1959, p. 76.

25. Ibid., pp. 75–76.

26. Egner and Denonn, *Basic Writings*, p. 255.

27. Kline, *Mathematical Thought*, p. 1192.

28. Ibid., p. 1195.

29. Egner and Denonn, *Basic Writings*, p. 255.

30. Clark, *Life of Russell*, p. 110.

31. Egner and Denonn, *Basic Writings*, p. 370.

Spherical Surface

1. Plato, *Timaeus and Critias*, trans. Desmond Lee, Penguin, London, 1965, pp. 45–46.

2. Bartlett, *Familiar Quotations,* p. 638.

3. Heath, *Thirteen Books of Euclid's Elements*, Vol. 3, p. 261.

4. T. L. Heath, ed., *The Works of Archimedes*, Dover, New York, 1953, p. 39.

5. Ibid., p. 1.

Trisection

1. John Fauvel and Jeremy Gray, eds., *The History of Mathematics: A Reader*, Macmillan, London, 1987, p. 209.

2. Cajori, *History of Mathematics*, p. 246.

3. Descartes, *Geometry of René Descartes*, pp. 216–219.

4. Cajori, *History of Mathematics*, p. 350.

5. P. L. Wantzel, "Recherches sur les moyens de reconnaitre si un Problème de Géométrie peut se résoudre avec la règle et le compas," *Journal de mathematiques pures et appliquees*, Vol. 2, 1837, pp. 366–372.

6. Ibid., p. 369.

7. Underwood Dudley, "What to Do When the Trisector Comes," *The Mathematical Intelligencer*, Vol. 5, No. 1, 1983, p. 21.

8. Robert C. Yates, *The Trisection Problem*, National Council of Teachers of Mathematics, Washington, DC, 1971, p. 57.

Utility

1. Kline, *Mathematics in Western Culture*, p. 13.

2. Richard Aldington, trans., *Letters of Voltaire and Frederick the Great*, George Routledge & Sons Ltd., London, 1927, pp. 382–383.

3. Kline, *Mathematical Thought,* p. 1052.

4. James Ramsey Ullman, ed., *Kingdom of Adventure: Everest*, William Sloane Publishers, New York, 1947, pp. 34–35.

5. René Taton and Curtis Wilson, eds., *The General History of Astronomy*, Vol. 2, Cambridge U. Press, New York, 1989, p. 107.

6. Albert Van Helden, *Measuring the Universe*, U. of Chicago Press, Chicago, 1985, p. 129.

7. Albert Van Helden, "Roemer's Speed of Light," *Journal for the History of Astronomy*, Vol. 14, 1983, pp. 137–141.

8. Taton and Wilson, *General History of Astronomy*, p. 154.

9. Ibid., p. 153.

10. Bartlett, *Familiar Quotations,* p. 275.

11. Willard F. Libby, *Radiocarbon Dating*, 2nd ed., U. of Chicago Press, Chicago, 1955, p. 5.

12. Ibid., p. 9.

13. Morris Kline, *Mathematics for Liberal Arts*, Addison-Wesley, Reading, MA, 1967, p. 546.

14. Hardy, *Mathematician's Apology*, p. 119.

15. Stillman Drake, trans., *Discoveries and Opinions of Galileo*, Doubleday, Garden City, NY, 1957, pp. 237–238.

Where Are the Women?

1. Nadya Aisenberg and Mona Harrington, *Women of Academe: Outsiders in the Sacred Grove*, U. of Massachusetts Press, Amherst, MA, 1988, p. 9.

2. Cecil Woodham Smith, *Florence Nightingale: 1820–1910*, McGraw-Hill, New York, 1951, p. 27.

3. Auguste Dick, *Emmy Noether*, trans. H. I. Blocher, Birkhäuser, Boston, 1981, p. 125.

4. Fauvel and Gray, *History of Mathematics*, p. 497.

5. Michael A. B. Deakin, "Women in Mathematics: Fact versus Fabulation," *Australian Mathematical Society Gazette*, Vol. 19, No. 5, 1992, p. 112.

6. "Earned Degrees Conferred by U.S. Institutions," *Chronicle of Higher Education*, June 2, 1993, p. A-25.

7. Virginia Woolf, *A Room of One's Own*, Harvest/HBJ Books, New York, 1989, p. 47.

8. Gillispie, *Dictionary of Scientific Biography*, essay on Sonya Kovalevsky, p. 477.

9. Koblitz, *Convergence of Lives,* p. 49.

10. Ibid., pp. 99–100.

11. Ibid., p. 136.

12. Albers et al., *More Mathematical People*, p. 280.

X-Y Plane

1. Descartes, *Geometry of René Descartes*, p. 2.

2. Whiteside, *Mathematical Papers of Isaac Newton*, Vol. 1, p. 6.

3. Descartes, *Geometry of René Descartes*, p. 10.

4. Ibid.

5. Boyer, *History of Analytic Geometry*, p. 138.

6. Albers et al., *More Mathematical People*, p. 278.

7. Boyer, *History of Analytic Geometry*, p. 75.

8. Ibid.

Z

1. Struik, *Source Book in Mathematics*, p. 67.

2. Ibid., p. 69.

3. Katz, *History of Mathematics,* p. 336.

4. Descartes, *Geometry of René Descartes*, p. 175.

5. Whiteside, *Mathematical Papers of Isaac Newton*, Vol. 5, p. 411.

6. Kline, *Mathematical Thought*, p. 254.

7. *A Century of Calculus*, Part I, Mathematical Association of America, Washington, DC, 1992, p. 8.

8. William Dunham, "Euler and the Fundamental Theorem of Algebra," *The College Mathematics Journal*, Vol. 22, No. 4, 1991, pp. 282–293.

Index